SAVING THE
PLANET WITHOUT
THE BULLSHIT

Assaad Razzouk is a Lebanese-British clean-energy entrepreneur, author, podcaster and commentator. He co-founded and runs a clean-energy company, headquartered in Singapore, financing, building and operating renewable energy projects in Asia. He also runs a not-for-profit Singapore start-up, digitizing and democratizing renewable energy. With his hands-on experience in renewable energy, combined with his other roles, Assaad is a high-profile thought leader on climate change, clean energy and the UN climate talks, with several hundred thousand followers on Twitter, Facebook and LinkedIn and occasional widely read newspaper columns.

SAVING THE PLANET WITHOUT THE BULLSHIT

WHAT THEY DON'T TELL YOU ABOUT THE CLIMATE CRISIS

ASSAAD RAZZOUK

Atlantic Books
London

First published in hardback and trade paperback in Great Britain in 2022
by Atlantic Books, an imprint of Atlantic Books Ltd.

10 9 8 7 6 5 4 3 2 1

A CIP catalogue record for this book is available from the British Library.

Hardback ISBN: 978 1 83895 462 8
Trade paperback ISBN: 978 1 83895 463 5
E-book ISBN: 978 1 83895 464 2

Printed in Great Britain by TJ Books Ltd

Atlantic Books
An imprint of Atlantic Books Ltd
Ormond House
26–27 Boswell Street
London
WC1N 3JZ

www.atlantic-books.co.uk

Non, ça ne me suffit pas

Contents

Introduction

It was the blackened teeth and poisoned skin of coal miners in China's coal capital, the city of Taiyuan, that gave me my first brush with sustainability. I had never seen anything like it before. The sight of those embarrassed smiles (instinctively covered with one hand) and awful complexions stopped me in my tracks and forced me to pause, think and question everything that I had until that point believed was important.

I first landed in Taiyuan in the spring of 2006. Flights there were unreliable and often cancelled. During the winter, chimney smoke enveloped the city. In the warmer seasons, the population was spared the smoke but it still had to contend with other scourges: pollution caused by coal-dependent industries, sand and dust from the surrounding countryside, slow-moving traffic and endless construction works as the city's unrelenting tentacles spread wider and wider. The air tasted gritty and the mountains on three sides helped trap air that looked yellow and felt like it was asphyxiating all those breathing it.

The coal mine and associated buildings lay to the east of the city. As I approached, the roads crackled, the heavy traffic became increasingly demonic and the air thicker. A sulphureous odour permeated the dusty atmosphere, generated by the poor-quality coal burned and breathed by the miners. The route to the mine itself was tortuous and hazardous. A narrow dirt track rose steeply up the side of a rubble-strewn valley and into the scrub-covered mountains. The buildings were dilapidated, and pavements were covered in dirt and debris thrown up by passing heavy trucks that also gouged ruts in the road. Groups of coal miners occasionally appeared as dark silhouettes in the dusty distance, as if rising from the land where their makeshift homes were buried.

I'd spent my formative school years in Lebanon, in the midst of a civil war where neither safety nor school schedules, roads, electricity or clean water were a given. After managing to get into a US university, I had no choice but to excel academically to land a job and therefore a work permit: the alternative was to return to a bitterly divided and burning Beirut. After university, it was all about competition and survival, first in the United States, then the United Kingdom and finally Singapore. I had little time to be aware of the environmental issues around me.

I was in China in 2006 on a business trip. At the time, I was trying to build a company that reduced quantifiable amounts of pollution from certain industries. We would then obtain proof that pollution had been reduced, in the form of certificates that we would sell into a market created especially for this purpose under a United Nations agreement called the Kyoto Protocol, a gallant effort focused on achieving net reductions in worldwide pollution. We also had to prove that we wouldn't have intervened to invest the necessary money and technology to decrease the pollution if it weren't for the Kyoto Protocol. The goal was to drive a sustained reduction in emissions from greenhouse gases everywhere.

Pollution is agnostic when it comes to its global impact. Climate change in particular is caused by emissions that are man-made: greenhouse gases, and pollution generated when we burn oil, gas or coal or anything made with them (like plastics). It doesn't matter where that pollution is generated; it all ends up in the same place: our planet's atmosphere.

In Taiyuan, we saw an opportunity to stop methane, a potent greenhouse gas around 85 times worse for the environment than carbon dioxide over 20 years, from leaking from Chinese coal mines in vast quantities. Our engineers helped build drainage pipes that captured the methane, prevented it from escaping into the atmosphere and redirected it instead to generate electricity.

However, what started as an abstract commercial opportunity became very real at the sight of the coal miners in Taiyuan, who had been visibly and permanently scarred by one of the unhealthiest activities in the world. It didn't take long to figure out that irrespective of how many vegetables they might eat, their health wasn't going to recover from mining and burning industrial quantities of hydrocarbons.

Over the next few years, I developed my business focusing on greenhouse gas mitigation in multiple countries, including the United States, China, India and South East Asia; then a business deploying renewable energy (solar and wind power projects as well as biogas to energy and biomass to energy projects) across Asia.

Over the course of 16 years (and counting) in carbon tech, clean tech and developing greenhouse gas mitigation and renewable energy projects from the ground up, I was at the front line of climate change-fuelled destruction. I witnessed the incredible vulnerability of Thailand, the Philippines, Indonesia, China, India and Pakistan to floods; the discernible warming trends in Asian cities; the plastic pandemic; the destruction of our oceans; the insect apocalypse going on in our midst; and the deforestation of the remaining rainforests in Indonesia.

I was also at the table at multiple United Nations climate meetings, sometimes as a government negotiator and other times as an observer or independent participant, as well as at a vast number of conferences convened by the United Nations, other governments and the private sector.

Every one of these meetings and every visit to vulnerable communities helped develop my awareness of the climate catastrophe in our midst. Over time, I started shouting more stridently from the rooftops about the climate change crisis, using social media, my podcast and occasionally newspaper articles and blogs. Throughout, I also tried to showcase solutions and actions we can all take to make a real impact.

The world has hosted 26 annual climate talks since 1995, at which practically every country has been represented. These talks were convened because human activities (principally burning fossil fuels) have dumped an enormous amount of greenhouse gases in the earth's atmosphere. This has had the effect of warming the planet to levels that increasingly threaten our survival, by shifting the frequency and violence of climate extremes such as droughts and typhoons, as well as causing the ice sheets to melt, the seas to rise and temperatures in some areas to become intolerable to humans.

During each of these annual climate talks, thousands at first, then tens of thousands of representatives from government, industry and civil society have held countless meetings, none of which bent the curve. Greenhouse gases continue to increase inexorably, and as they do, promises of decarbonization recede further and further into the future. What is, however, sometimes less appreciated is that a very slow decarbonization path is a set-up for much more absolute harm than a very fast one. We can think about this the way we think about a mortgage loan: if you reduce the capital of the mortgage quickly (in the case of climate change, the capital would be total greenhouse gas emissions), you would save a lot in interest (i.e., climate effects), and as a result, the overall amount you pay out is greatly reduced. Every fraction of a degree of warming matters. The 2015 Paris Agreement agreed by 192 countries is aimed at limiting the average increase in global temperatures to well below 2° Celsius above pre-industrial levels, and preferably to 1.5°. The warming implications of 1.5° versus 2° Celsius are very different. For example, if rapid decarbonization limits warming to 1.5°, then coral reefs around the world would be projected to decline by 70 per cent, whereas at a 2° increase, they would decline by more than 99 per cent. In a 1.5° warming scenario, heatwaves will affect 14 per cent of the world's population every five years, whereas 37 per cent (that's close to three billion people) would be affected in a 2° world.

Yet we must not just talk about failures, as we seem to do – overwhelmingly – in the climate change space. Yes, we have failed to control emissions or to even bend their upward-sloping curve.

Yes, extreme weather and climate disasters are now a fact of daily life for many. But every moment spent dwelling on these setbacks is a moment lost to effecting change. What's required is a focus on strategies in our daily lives that deliver outcomes.

While informed individual efforts to be eco-friendly are important in galvanizing others, this book argues that they largely miss the mark. I'll show that individual action, while good and important from a moral standpoint, makes little actual difference and may even be counterproductive in some cases. We are in critical need of major systemic changes: 89 per cent of emissions come from burning oil, gas and coal, and just stopping subsidies for fossil fuels would cut global emissions by a third. Yet cash subsidies for fossil fuel consumption alone amounted to $440 billion globally in 2021. Once we include all assistance to the industry, we paid out $11 million every minute, according to no lesser authority than the International Monetary Fund. All this to burn up our planet faster.

How do we make these systemic changes happen? And what is the most effective role for individuals as we switch to clean energy to fuel global lifestyles?

All over the world, people are confronted with complex daily choices in the name of saving the planet: can I avoid using this plastic bag? Should I fly less? Should I get an electric bike? Am I buying too many clothes? Are electric cars *really* good news? Should I bother to recycle? Is it ethical to eat my favourite burger or must I go vegetarian?

This book is an attempt to clear a path through all the clutter surrounding our daily efforts to do the right thing, while also trying to care for our loved ones and to survive and thrive.

It is structured to cover large segments of what we all do on a day-to-day basis, including eating and drinking, going about our daily lives, travelling and investing. In each chapter, I try to debunk the myths peddled by many – including lobbyists for oil companies and for the status quo, but also by well-meaning citizens. I seek to replace these with a dose of common sense in order to put together a manifesto of how to live in the midst of a climate catastrophe and how to effect real change. As citizen consumers, we are scraping together mere pennies reducing, reusing and recycling (for example), while oil companies burn hundred-dollar bills.

A recent social media stunt by oil giant Shell illustrates this. In November 2020, the company asked the public via a poll on Twitter: 'What are you willing to change to help reduce emissions?' The campaign backfired spectacularly when hundreds of thousands pushed back, some responding with their willingness to hold Shell accountable for obfuscating the truth about their products and their impact on the environment for over 30 years when it secretly knew the entire time that fossil fuel emissions would destroy our habitat.

The fact that Shell, after all that has happened in the last few decades, *still* has the gall to try and mislead the public shows that the road ahead continues to be paved with the bad intentions of wealthy and destructive corporations. Regulators can barely keep up. In the Netherlands, for example, the Dutch Advertising Code Committee ruled that Shell was misleading automobile drivers in its advertising campaign of 2020 and 2021 that told them that if they paid an extra cent per litre of fuel, they would be driving in a carbon-neutral way (in reality, we don't know they are until every tree planted to offset their emissions – if in fact it was planted at all, which isn't always the case – has survived decades). But this was a single advertising campaign and regulatory action was only possible because nine students filed a legal complaint backed by climate campaigning organizations. Civil society doesn't have the resources to keep up with the torrent of greenwashing all around us and take action, because Shell is not alone. Since the 2015 Paris Agreement, and in spite of numerous announcements about how they were committed to fighting climate change, Exxon, Shell, Chevron, BP and TotalEnergies collectively spent $200 million a year on lobbying to expand fossil fuel operations, and more on advertising.

Enough is enough. It's time to change the conversation.

1

Plastic Is Your New Diet

There are scientific papers with titles like 'The human consumption of microplastics'. Wait, *what*? Surely we can all agree that plastic isn't food and that we shouldn't be eating, drinking or breathing it? Maybe not. Right now, worldwide, we eat, drink and breathe plastic with every meal, every drink and every breath we take. We do so at the rate of about 200,000 plastic particles for every human being on the planet every year.

The word 'plastic' is manipulative in the extreme. Plastic's building blocks, called polymers, are in 99 per cent of cases made from coal, gas and crude oil. It's a very comfortable word, however, not least because advertising campaigns over decades have extolled its virtues. It definitely beats 'made from oil' as a label. Imagine if consumers worldwide looked at fast fashion, for example, and instead of 'made from polyester', they saw 'made of plastic derived from crude oil and produced using harmful chemicals, including carcinogens'? *That* would dampen sales. It's a tour de force of branding.

Microplastics are generated when common household items like bags, clothes and cosmetics made partly or wholly from plastic break into tiny particles and enter our environment. These particles then circulate and slowly, inexorably make their way into our rivers, our oceans and our groundwater before coming back to us via our drinking water, our vegetables, our fish, our meat and our air.

Microplastics are found in over 83 per cent of the world's tap and bottled drinking water. Not surprisingly, they've been also confirmed in the placenta, in newborn babies and in children, and in 2022, in our blood and organs.

We have contaminated everything with plastic to the extent that periodically, dead whales wash up ashore with up to 40 kilos of plastic in their bellies. They die of dehydration and starvation after eating plastic fishing nets, plastic rice sacks, grocery bags, banana plantation bags and general-use plastic bags that have been thrown into the ocean.

It's a sustained carpet-bombing campaign. In the pristine Pyrenees mountains of France and Spain, microplastics land at the rate of 365 particles on every square metre every day. New York? Paris? London? Karachi? Beijing? Delhi? Cairo? They're coming down at the same rate. We've also discovered plastic particles in the rain in the Rocky Mountains, and even in the falling snow in the Arctic. I don't know about you, but I dislike intensely the thought of eating, drinking and breathing plastic in order to line the pockets of those who make and sell the stuff (or, more accurately, force it down our throats).

The companies that make microplastics are the same ones responsible for the climate change emergency: the oil, gas and petrochemicals companies. They are enjoying a free ride by dumping plastic on us with little thought or control, in collaboration with some of the biggest companies in the world: Coca-Cola, PepsiCo, Nestlé, Danone, Kraft, Procter & Gamble, Unilever, Mars, Colgate-Palmolive, all brands that make loud claims on their websites about how much they care about nature, sustainability and the environment.

These companies are wild about plastic principally because it's mispriced. The oil, gas and petrochemicals companies make money from what is otherwise a waste product from their principal activity (extracting and burning oil and gas). They're not paying for any of the damage caused by plastics, and neither are consumer companies like Coca-Cola or PepsiCo or Nestlé or Unilever, which use plastic bottles and plastic containers with minimal restraint or respect for the environment. As a result, they can all sell their products far cheaper than they should, and make money where they shouldn't. It really is an extraordinary gig: unleash poisonous pollutants everywhere, completely free of charge, and make lots of money doing it.

If the destructive impact of producing plastic was priced into products, society would not be able to afford it. Most single-use plastic would disappear relatively quickly (after we'd figured out how to transport, distribute and sell products that depend on plastic for their shelf lives), replaced by a myriad of alternatives that would become cost-competitive, such as plastic from recycled polymers. Unfortunately, today it's cheaper to buy new plastic than to pay someone to manage and sort recyclable plastic, which is a tiny proportion of the massive amount we produce.

We all have a role to play in tackling plastic pollution, but right now, individual consumers are the ones carrying the burden. There's only so much that citizens can do if companies don't step up and provide sustainable choices and governments don't tax or regulate plastic out of the system. We don't need at least 90 per cent of it – we have substitutes – but here again, individuals can't get the job done on their own.

Take bioplastics, which could be used for products ranging from food packaging to medical devices. Bioplastics are plant-based plastics made from sugar extracted from corn and sugar cane; grown from algae; or engineered from scratch using bacteria. I am sure most of us would love to replace the plastic products we use with their bioplastic equivalents, but that would work only if local and central governments invested in the infrastructure required to recycle them: most bioplastics don't break down naturally and require specialized composting facilities.

The data are shocking. In 1973, each person on earth used two kilos of plastic a year. Fast forward to today, and each one of us uses 46 kilos of plastic. That's 23 times more, mostly for no good reason. We can think about these 46 kilos of plastic in a different way: the amount of plastic produced *in one year* is roughly the same as the entire weight of all 7.7 billion people on earth. Since we invented the stuff, we've produced the equivalent weight of 100 billion people. That's right. One hundred billion people.

How could we possibly need the one million plastic bottles bought worldwide *every single minute*? These bottles last 450 years when they make their way into the oceans, which they do all too often these days. Once there, they kill 100,000 marine mammals and turtles each year, because these animals can't distinguish plastics from food. Scientific studies have documented that so far, 701 marine species have ingested plastic, while 354 species have been seen entangled in it.

Back on dry land, at least 400 different species of animals have either ingested plastics or become entangled in them. The microplastic then makes its way back into our food chain, our drinking water and the air we breathe. So we're eating plastic. We're drinking plastic. And now we also know that we're breathing plastic. But *why*?

First, it's because we don't actually recycle plastic. Plastic manufacturers have convinced the public that plastic pollution is not their problem, because their product can be recycled. Unfortunately, plastic recycling is a sophisticated con, a device used by manufacturers and distributors to make us feel good about what they are selling us – and therefore sell us even more of it. In reality, *there is no plastic recycling infrastructure at scale anywhere in the world*. We just assume there is because those who make and sell plastic talk about recycling, and because they made sure that this pretend recycling makes us feel virtuous. In fact, most of the plastic we use in our daily lives is either not recyclable or no business bothers recycling it.

We've recycled less than 7 per cent of the plastic we've ever produced, and at least another 5 per cent has been dumped into the oceans. That's equivalent to the weight of five billion people. The rest? It's either burned in the countries of consumption or shipped to developing countries in Asia such as the Philippines or Thailand or one of several African countries, all of which lack recycling infrastructure, most likely to be burned or left in landfill.

Recently, the desperation of the we-just-pretend-to-recycle-plastic industry was such that a US industry lobbying group, the innocuously named American Chemistry Council (a front for some of the world's largest oil, gas and petrochemical companies), was pushing to ensure that an emerging trade deal between the United States and Kenya required Kenya to *reverse* its strict limits on plastics (Kenya bans single-use plastic) and to continue to import plastic trash from the US. Their ambition was to flood Africa with plastic garbage, using Kenya as the bridge.

Thankfully, pretty much all of the places mentioned above are now trying to stop countries like the US and the UK dumping trash on them, and hopefully Kenya will do so too.

The pretend-recycling brouhaha is in reality only about recycling stuff that allows factories to make money. For example, high-value plastics like fizzy drinks bottles, which come stamped with a number 1 on them if you look closely, are probably recycled. Stuff stamped with a number 2, like milk cartons or shampoo bottles, *may* be getting recycled, at least some of the time. Everything else, stamped from 3 to 7, is neither sorted nor recycled.

There's another problem. Consumer goods companies don't always want to use recycled plastic. That's because new plastic is a lot cheaper: its manufacturers don't have to invest in the infrastructure needed to recycle it, nor pay for the climate and pollution and environmental damage they're causing.

We eat, drink and breathe plastic because it's a waste product of the oil and gas industry and because of the obscene money that has been made available to petrochemical companies to manufacture a lot more of it, insanely cheaply. Plastic is made from either ethane, derived from natural gas, or naphtha, which is distilled from crude oil. All that stuff goes into multi-billion-dollar petrochemical plants (note that word 'petro': these are chemicals derived from petroleum or natural gas), and out comes plastic.

Petrochemical plants are typically owned by mega corporations; for the most part, the same cast of characters responsible for 70 per cent of all global emissions since industrialization began, led by companies like ExxonMobil, Saudi Aramco, Chevron, Shell, ConocoPhillips, and Chinese companies like CNPC and Sinopec. If you think that each of us using 46 kilos of plastic each year is a lot, you'd be right. But the leadership of these companies are trying to multiply that number by 40 per cent over the next five years.

Since 2012, oil, natural gas and chemical companies have poured around $200 billion into 300 petrochemical plants across the United States. There are another 108 petrochemical plants scheduled to come online in Asia and the Middle East by 2023. All these exist to do only one thing: sell us more milk cartons, shampoo bottles, supermarket bags, straws and food packaging.

For a while, this boom was driven by fracking. Particularly popular in the United States, fracking involves pumping an enormous amount of mostly unregulated chemicals into the ground, along with sand and water, in order to provoke gas to gush out. Then we grab it. This activity resulted in the US in an over-abundance of fossil gas, which collapsed prices, giving American petrochemical companies a huge advantage over foreign producers, but only if they could build new plants and expand market share. Which is exactly what they're doing. Low natural gas prices also made it economical to liquefy that gas and ship it around the world.

To add oil to the fire, these bad actors have easy access to vast amounts of capital. The fracking industry loses billions each year, yet in the last few years, petrochemical companies have had no problem financing $300 billion of new kit. And that's despite the global push-back against plastic, and even though the banks and capital markets know precisely how incredibly destructive it is.

The formula of the US shale fracking industry is to borrow billions of dollars at low interest rates, lose an enormous amount of money, force oil and gas from marginal fields, and then leave the public stuck with the financial losses and wanton environmental destruction. After all, someone has to pay to clean up all these abandoned oil and gas wells. According to the US government, there are three million of them already. These fracking companies have a cash-flow deficit of $20 billion per year, which means they make $20 billion per year *less* than their operating costs and capital expenditure. And this has been happening for a long time, with limited exceptions when oil and gas prices spike.

Why does the fracking industry and therefore the petrochemicals industry continue to receive investments from Wall Street despite rarely making any money? Wells Fargo, J. P. Morgan, Citi, Barclays, Bank of America and Goldman Sachs are all in on it. The answer lies in the sheer size of the sums involved. Petrochemicals and fracking are big-numbers businesses, and that's how the banks make money. Whether the fracking companies are profitable or not doesn't matter to the bankers. They're getting rich in the short term, making huge loans to an industry that likely cannot pay them back in the long term. That's because bankers are incentivized, through their bonus systems, to care over the 12-month period in which the bonus is calculated. If a company they backed goes belly-up three or four years after they've lent them the money, they probably won't even be around to see it.

ExxonMobil, Shell, Chevron, Saudi Aramco and others are going all in on petrochemicals because they know that demand for oil in the transportation sector is going down, and they've promised their shareholders that they're going to keep demand at the level it's at today – and even increase it – by flooding the world with more and more plastic. In their forecasts, the International Energy Agency, BP and ExxonMobil expect 2 to 4 per cent annual growth rates for plastic demand, from the stratospheric levels it is at today, which means a doubling of demand in 18–24 years. That's what the industry is tooling up for.

But they know, and we know, that people around the world want to reduce their hydrocarbon usage, whether it's fuel for cars and buses, or plastic. Will they succeed?

I don't think so. Single-use plastic is under attack in over 40 countries, either through surcharges, restrictions or outright bans. What we really need is a universal ban.

Baby steps are under way to agree a proposed international treaty on plastics. In 2022, the United Nations Environment Assembly adopted a resolution calling for a legally binding treaty to combat plastic pollution. This still needs to be drafted and agreed, but with luck, it could be in place later this decade to help end the flow of plastic into nature and hold polluters to account.

Meanwhile, all countries should do what the European Union, Canada, Kenya and the city of Mumbai all mustered the courage to do: ban single-use plastic everywhere.

It's not that difficult. An unexpected global trailblazer in the area of plastic bans is Rwanda. Despite still recovering from genocide and war, the African country has nonetheless outlawed the import, production, use and sale of plastic bags and packaging since 2008, with exceptions for specific industries such as hospitals and pharmaceuticals. This ban makes eminent sense. Governments everywhere, especially in poorer countries or smaller cities, lack the funds to build resilient sanitation and an effective recycling infrastructure to deal with mountains of plastic trash. Often plastic bags are burned instead, releasing toxic pollutants into the air and contributing to clogging waterways, pipes and drains, sometimes leading to devastating floods.

The contrast between Rwanda and Singapore, one of the richest countries in the world, is stark. I recently purchased cut papaya fruits in Singapore, which came in a plastic bag inside a plastic container that checkout staff subsequently tried to place in another small plastic bag. Surely such a highly organized country could ban single-use plastic overnight?

Other single-use plastic items like straws, coffee cups and cutlery also need to be replaced by biodegradable ones. Alongside simply banning single-use plastic, we also need to tax all other types, to stop them being dumped on us (in the form of fast fashion, for example). We need to price plastic *out* by pricing *in* its pollution. Only national governments can do this.

It's also critically important to ensure that petrochemical companies invest in recycling infrastructure. All plastic producers should be required to provide the equipment required to recycle plastic or dispose of it properly. If that happened, most plastic would become uneconomic because its price would increase. High-end plastic – the stuff we need for medical equipment, for example – would remain, but the pollution from non-recyclable or un-recycled plastic would completely stop: recycling only happens when the plastic itself is recyclable (which most of it is not) or when companies agree to pay for recycled plastic (which is not the case for most plastic products, including fast fashion). Put another way: if you can't recycle it, you shouldn't be allowed to produce it.

Recycle plastic if you want, but when you do, bear in mind that if the plastics industry recycled even *half* as much as the aluminium industry, *we would never need any new plastic*. Ever. Recycling is completely against the interests of the petrochemical giants, which can only increase revenue and profit by selling more. Entirely new industries are growing that offer consumers an alternative. Support them with your spending. Find these new industries and invest in them. They are the future.

Limiting plastic demand is great, because this would devastate the oil and gas industry's plans to bury us in garbage and slash demand for oil and natural gas. As a bonus, it would also make a huge dent in greenhouse gas emissions. Despite all the talk about climate, despite climate change conferences going back nearly 30 years, emissions continue to rise and we have still not yet turned the corner, as we must: we have until 2030 at the latest for emissions to start dramatically declining.

Local governments too, inundated with plastic waste, are growing desperate. They are doing what they can, but again they are focused on the pollution, not its causes. In two countries very far apart, Indonesia in South East Asia and Ecuador in Latin America, local governments, threatened by uncontrolled amounts of plastic, introduced innovations to help their poorest inhabitants while trying to do something about the scourge. The cities of Surabaya in Indonesia and Guayaquil in Ecuador, both with around 2.8 million people, introduced schemes to pay your bus fare with plastic bottles.

The queues for buses in Surabaya were a sight to behold. People could ride for an hour on public buses with an unlimited number of stops and could pay to do so with either three large plastic bottles, five medium plastic bottles or ten plastic cups. In Guayaquil, 15 plastic bottles bought you a ticket on the city's bus transit system. But then what? The cities collected lots of plastic trash, which they can't do much with. They either burn it, worsening air pollution, or the plastic finds its way to the ocean.

Ultimately, it is dramatically reducing the *supply* of plastic, not its demand, that will make the biggest difference.

If oil companies paid for the environmental impact of their products, or if petrochemical companies were obliged to invest in recycling infrastructure, plastic would be a great deal more expensive. Governments must step in to accurately price its environmental impact. It's imperative that we all make our voices heard in order for things to change.

See also:
Chapter 9: Stinky Gas;
Chapter 23: The ESG Con

2

Who Put Palm Oil in
My Toothpaste?

There's a tree called the oil palm tree and it has a magical fruit. If you squeeze the fruit, it produces a very special kind of oil called palm oil, which we've used for thousands of years. Originating from West Africa, palm oil was already a staple food 5,000 years ago. It was ascribed such a high value in ancient societies that we find it in casks alongside people buried in tombs at Abydos, a city in pharaonic Egypt, and in the royal cemetery of the pharaohs of the First and Second Dynasties.

Palm oil does an amazing number of things. It makes cookies healthier; it makes soap bubblier; it makes crisps crispier; it makes lipstick smoother; it even keeps ice cream from melting. Palm oil is in an astonishing 50 per cent of the packaged products that we find in a supermarket. It's in everything from pizza to doughnuts; chocolate to deodorant; shampoo to toothpaste; lipstick to animal feed. It's even used as biofuel. It's endowed with amazing advantages.

Palm oil is semi-solid at room temperature, which means it can be used to make spreads and keep them spreadable. It resists oxidation, which gives products a longer shelf life. It's stable at high temperatures, which means it keeps chicken nuggets crispy and crunchy. It doesn't have any smell or colour, which means it doesn't affect the look or taste of food products. It provides the foaming agent in every shampoo, liquid soap or detergent you can think of. It's truly a magical substance. For these reasons, the reach and footprint of palm oil is amazing. It uses 9 per cent of global cropland devoted to oil production. Every year, 3 billion people in 150 countries consume 8 to 10 kilograms each of the stuff. The best part? Almost no one appears to know how ubiquitous it is in our everyday lives.

Palm oil production at scale arrived in South East Asia in 1917, when the first commercial palm oil plantation was established in Malaysia by Frenchman Henri Fauconnier, a young entrepreneur seeking to build his fortune in the Far East, like many other Europeans at the time. Fauconnier had seen the increased demand for palm oil after the Industrial Revolution, because of its versatility and the ability to use it in multiple products such as soap, candles and industrial lubricants.

Today, two Asian countries supply 85 per cent of the world's palm oil: Indonesia and Malaysia. Another 42 countries also produce it, albeit in much smaller quantities. Under the hood, however, there are very serious problems with palm oil, its magical properties notwithstanding. First of all, a lot of palm oil companies have been burning rainforests in order to plant oil palm trees. The scale is incredible. While cattle ranches and soybeans drive Amazon deforestation, palm oil plantations win the deforestation prize in Indonesia. Deforestation alone pushed Indonesia into the top tier of global emitters of greenhouse gases powering global warming, alongside the US and China, just because rogue palm oil companies are torching rainforests to replace them with palm oil trees.

Neighbouring countries are also affected and periodically asphyxiated by the haze that pollutes cities like Singapore and Malaysia's Johor Bahru, where the health impact costs in the tens of billions of dollars. Palm oil companies are also killing wildlife. Half of the Bornean orang-utan population has been wiped out in two decades because the palm oil industry is destroying its habitat. Elephants are also being annihilated, and another 193 species are classified as critically endangered or vulnerable, primarily as a result of palm oil production. Palm oil companies also have a documented track record of exploiting workers, children and local communities.

Despite all these problems, we won't switch to alternative vegetable oils because palm oil is incredibly efficient. If we stopped using it, we would need 4 to 10 times more land to produce substitutes, and in the process, we would threaten other habitats and species. For example, soybean oil – a substitute vegetable oil and the second most important oil crop in the world after palm – requires six times more land per unit of oil, and its cultivation is linked with harmful biodiversity and deforestation impacts in Argentina and Brazil. In addition, palm oil creates a lot of jobs in Indonesia, Malaysia and elsewhere. Millions of smallholder farmers depend on it, and the wages it generates feed a huge number of people.

For these reasons, boycotting palm oil does not work, even if a typical consumer knew which products to avoid in a boycott. It also wouldn't necessarily be effective, because 40 per cent of palm oil is consumed in China, India and Indonesia, where Western consumer boycotts make little impact. India has seen the fastest growth in consumption. We can expect that to continue, which means that in all likelihood we'll need more palm oil, not less.

What we're doing with palm oil is not remotely sustainable. It's a scandal, but it doesn't have to be this way. However, the solution is not in the hands of consumers: effective climate solutions can only come via systemic change, and citizens shouldn't waste energy on ineffective actions unless these make them feel better, in which case that's just fine! Practical, implementable and comprehensive solutions are available, and that's where our efforts need to be focused.

Number one, we need palm oil production to be sustainable. This means that big buyers, not only those in the West, need to purchase the product only from sustainable sources, by tracking where their palm oil comes from and refusing any that is demonstrably linked to deforestation, human rights abuses and wildlife destruction. But that's hard to police or monitor.

The second solution is much more powerful. Some of the biggest users of palm oil are large multinationals like Colgate-Palmolive (soaps, toothpastes, deodorants and house cleaners, among other products), General Mills Hershey (chocolate), Kellogg's (cereals), L'Oréal (cosmetics), Mars (chocolate), Mondelez (chocolate), Nestlé (chocolate), PepsiCo (snacks) and Unilever (soups, ice creams, home and personal care products).

Some of these companies *are* trying to do something in terms of ensuring their palm oil is from sustainable sources, but there's a lot of pretend-doing going on because of a lack of accountability: directors who sit on their boards are not personally liable for environmental destruction. If the directors *were* held legally responsible for the environmental harm caused by their supply chains and as a result insurance companies stopped covering environmental destruction in their policies, everything would change overnight. Suddenly palm oil would become sustainable.

Moves are afoot to make this a reality. In the European Union, for example, the executive branch has already tabled a law requiring companies importing soy, beef, palm oil, wood, cocoa, coffee and certain derived products such as leather, chocolate and furniture to conduct mandatory human rights, sustainability, forest and eco-system risk due diligence to ensure their imports into the region are not linked to deforestation. The proposed law has teeth, imposing penalties of up to 4 per cent of sales on companies found to be non-compliant. In other words, companies will be forced by law to ensure that their products (and how these are procured), don't capitalize on human rights abuses or environmental destruction on the way to the consumer. When the law is adopted, member states of the European Union will subsequently implement a civil liability regime to give effect to the legislation.

This type of legislation already exists at individual country level in selected jurisdictions such as France and Germany and is making its way through the UK Parliament as well. Because of its nascent nature, it will need to be tested in the courts before its boundaries are established, including its precise impact on directors' liabilities. This is under way currently in France, for example. Finally, a draft UN Treaty on Business and Human Rights has been mooted since 2014, and while six rounds of discussions and negotiations have already taken place, it is in desperate need of political momentum.

Great strides in fighting deforestation can be accomplished if these efforts were given a strong push by direct citizen action modelled on Extinction Rebellion, the non-violent global environmental movement using civil disobedience actions to force governments to do the right thing on climate. Laws must be enacted everywhere to require companies to clean up their supply chains, by holding them responsible for knowing where they get their palm oil from (as well as other parts and products they use in what they sell consumers) and ensuring their suppliers aren't violating human rights, labour and environmental standards applicable in the company's country of incorporation.

For example, the directors of a German company selling products containing palm oil would be liable under German law for establishing that their Indonesian palm oil suppliers were not destroying rainforests to source that palm oil, abusing their workers, employing child labour or displacing indigenous people from their ancestral homes. Such laws would force these companies to conduct effective due diligence on their suppliers worldwide, pretty much instantly solving the problem associated with palm oil, because senior executives running multinationals would no longer be able to boost their companies' profits and reward themselves with fat salaries while taking shortcuts that cost lives and accelerate deforestation.

Analysis by the Commonwealth Climate Law Initiative, a UK-based non-governmental organization focused on analysing board directors' legal obligations regarding climate change, shows that company directors and officers in more than 21 countries could be held liable for failure to address climate-related risks. Non-governmental climate action organizations such as ClientEarth, an environmental law charity, are also taking action. They are trying to enable plaintiffs to begin pursuing directors for breaches of fiduciary duties (the legal obligations on directors to act in the best interest of their stakeholders, including shareholders and society at large) in cases where they are found not to be addressing climate risks, including deforestation risks.

Financiers backing climate-destroying corporations shouldn't be spared. Bank directors, as well as the trustees of pension funds and directors of investment management companies, should be held liable for harm to biodiversity caused by their funding. Research into the deforestation activities of 200 major banks and institutional investors showed that they had backed companies linked to deforestation to the tune of $279 billion since the adoption of the Paris Agreement in 2015, overwhelmingly without robust due diligence procedures related to human rights, labour and deforestation impacts. Yet these banks and institutional investors (and their directors) suffer no adverse consequences for now, while reaping rich profits from the resulting climate destruction.

Sometimes complex global problems have easy solutions.

See also:
Chapter 22: Tinker, Lawyer, Banker, Fry;
Chapter 27: Sue the Bastards

3

The Fashion Show at the End of the World

Every time you buy a piece of clothing, you should remember that the fashion industry is an environmental horror show. Yet the thought rarely, if ever, crosses people's minds. That's because fashion brands – from Gucci, Louis Vuitton and Prada to the more affordable Zara or Gap – are highly adept at manipulating our feelings. They rarely use taglines or slogans. When they advertise, they target your emotions with images and simple phrases. You imagine yourself wearing the $1,500 dress and carrying the $800 bag. You are transported to the fancy hotel, the perfect beach where the beautiful people are photographed.

The industry does a brilliant job positioning itself outside of rational questions over sustainability or environmental impact. It's with *feelings*, rather than reason, that you buy a T-shirt, a sweater, a handbag, trousers or one more pair of Nike trainers. Yet fashion's manufacturing processes release vast amounts of climate-warming emissions. In all likelihood, your new item of clothing also releases microplastics every time it's washed, which then make their way into the food chain. When discarded (as most apparel is), it's either burned or it decomposes into yet more microplastic.

In 2019, an online fast fashion brand (aptly named Missguided) released a bikini for $1.30. That's *a dollar and 30 cents* for an entire bikini set. It was made from 85 per cent polyester. Polyester is not biodegradable and takes at least 200 years to decompose (some scientists think it will never do so).

Someone bought this dollar bikini and wore it on a pristine beach. The bikini inevitably fell apart because of its quality. Its owner then threw it away and the polyester in it eventually came back to destroy that same beach. In general, fewer than 1 per cent of the materials consumed by the fashion industry go back into the system. The rest ends up in landfill or the sea, or are burned.

The fashion industry produces a mind-boggling 100 *billion* new items of clothing every year. Almost all of them are made from plastic, and with pretty much zero thought about where they'll end up. Over the past 15 years, clothing production has doubled, while at the same time we're using our clothes 36 per cent less and the number of garments purchased each year by the average consumer has gone up 60 per cent. We're producing a lot more clothes because we're buying a lot more clothes, which means that we're throwing away a lot more clothes as well. The data is scary, and fixing fast fashion appears daunting at first sight.

Meanwhile, reports routinely reveal that the fashion industry's efforts to mitigate its impact on the environment and improve its scorecard on social issues (such as living wages for garment workers) are critically insufficient.

You might think that fashion brands, which spend a great deal of time worrying about how they appear to us, the consumers, would be thinking about their materials; adopting a sustainable economic model by paying attention to how they manufacture their garments so that they can be deconstructed to be part of a circular economy and embracing recycling. You might also think they would want to be powered by clean, renewable energy. But the truth is, very few of them do much of this at all. As a result, the industry contributes to somewhere between 8 and 10 per cent of global greenhouse gas emissions, which is more than aviation and shipping combined. It's a major contributor to climate change because of its long supply chain and its energy-intensive production methods.

In addition, the fashion industry exacerbates global water scarcity by monopolizing 3 per cent of all the fresh water used for irrigation to produce cotton. It also produces 20 per cent of global wastewater (because of how much water it needs to treat textiles and dye them). This is dirty water full of chemical compounds that is dumped into rivers and oceans, often in a wholly unregulated manner. Fast fashion uses, incredibly, more than 15,000 different chemicals, and no comprehensive estimates exist of how they impact on human health. Finally, we – and the industry – throw away enormous numbers of clothes, some of which find their way to our rivers and oceans and decompose into the toxic microplastics that we ingest. Fast fashion contributes approximately 35 per cent of the microplastic pollution we add into the oceans each year.

The Fashion Transparency Index (FTI), a tool to incentivize and push major brands to disclose more information about their policies, makes very illuminating reading. The FTI analyses 250 major brands, each with revenues of more than $400 million a year;

it reviews and benchmarks their public disclosure on human rights and environmental issues across several key areas, then scores them. Areas covered include what the brands disclose about their policy and commitments; their governance system and its robustness; the traceability of their supply; how much work they do to monitor and assess their suppliers and their practices; and how they manage issues including their COVID-19 response, their waste, their water usage, chemicals, climate change and deforestation.

What emerges year after year (the index has been published since 2017) is shocking. The 2021 FTI showed that with just a single exception (Venice-headquartered Italian clothing company OVS), every brand scored below 70 per cent transparency, and 20 brands scored a flat zero. In other words, 249 brands disclosed too little for the consumer (or anyone else, for that matter) to understand what they're doing or how they're doing it; have little in the way of policies regarding hiring practices or local community engagement; don't assess their suppliers; don't have processes to understand or stop environmental damage; don't care how they make their garments, and don't care what happens to them after they are thrown away.

The better performers (24 brands scored 50 per cent transparency or more) include household names Adidas, H&M, Nike, Patagonia, Reebok, Gucci, C&A and Vans. All of these are trying, but can do so much better. Those with a transparency score between zero and 20 per cent include sports and outdoors brands Sports Direct, Fila, Billabong and Quiksilver; department stores Macy's, Bloomingdale's and Saks Fifth Avenue; and even luxury brands Chanel, Brunello Cucinelli, Dolce & Gabbana, Prada, and Armani. It's a very long list of 130 major brands distinguishing themselves with appalling disclosure standards and little or no commitment to fighting environmental damage, implementing safe hiring practices or paying fair wages.

Inconsistencies and gaps are rife. For example, more than 60 per cent of the world's clothes are made from synthetic fibres; in other words, oil. (Cotton is the next biggest source, accounting for over 30 per cent, followed by natural fibres originating from plants at 6 per cent and wool at less than 2 per cent.) Because oil powers fashion, the industry is a large contributor to climate change, but also to our waste management issues: your (plastic) T-shirt is not biodegradable, and requires a substantial amount of chemicals to manufacture, emitting toxic pollutants along the way and releasing tons of microplastics. Yet only 30 per cent of the brands even provide a definition of what constitutes a sustainable material, so how can we trust them to manage what they can't even define? Furthermore, only 25 per cent publish robust targets on cutting the use of textiles made from oil.

In the 'good news' column, some companies *are* forming coalitions to try and deal with the environmental and social challenges of fashion. To tackle carbon emissions, for example, approximately 130 global brands – including Lululemon, Adidas, Nike, LVMH and Guess – have joined the Science Based Targets initiative and must in time make and stick to commitments to use renewable energy across their entire business, including their supply chain.

Another 160 stakeholders in the fashion industry – many the same as those committed to cut carbon emissions, including Adidas, Burberry, H&M, LVMH, Nike and others – belong to a coalition initially called Zero Discharge of Hazardous

Chemicals (just the name gives a sense of how much we don't know about fashion) and now rebranded as the Roadmap to Zero Programme, which aims to reduce the industry's chemical footprint.

Good examples are few, however, while bad examples are everywhere. Dealing properly with hazardous chemicals should be the rule, not the exception. Approaching cotton sustainably or cutting carbon emissions, similarly, should be the default position, not a goal to strive for.

It's very hard to expect consumers to figure out which brand is doing what before they head out to shop, then keep the pros and cons in mind when they buy an item. That's because while individual action is great, the fashion landscape is confusing and complicated and the time required to properly analyse every purchase from a sustainability perspective (assuming there are clear answers to be had) clashes with the instant gratification we associate with shopping. Researching the environmental footprint of a brand might be a great idea, but it is difficult.

Take as an example the modal fabric manufactured by an Austrian company, Lenzing AG. Modal is a semi-synthetic fabric made from the pulp of beech trees combined with other fibres. It was created in Japan in 1951 as an alternative to silk that is vegan, strong, breathable and luxurious. Headquartered in Lenzing, a small town of 5,000 people halfway between Vienna and Munich, Lenzing AG uses certified sustainable wood, recycles its water, reuses almost all of its solvent, discloses the entire carbon footprint of its supply chain and is one of the fashion world's rarest breeds: a company with a circular business model that wastes almost nothing and recycles almost everything. Furthermore, modal is easily biodegradable and compostable, uses 20 times less water than cotton, while its yield when grown can be 10 times higher. Its advantages don't end there: when washed, it doesn't release microplastics and doesn't need either whitening agents or fabric softener. Finally, it's luxuriously comfortable – like silk – and doesn't wear out easily.

Who knew, right?

But what's a consumer supposed to do when shopping? We now know how good Lenzing AG are on sustainability, but what about all other modal fabric manufacturers? Should we buy their products, or would we be accelerating deforestation if we did?

Should we check every label to determine how much plastic is in the shirt we want to buy? Which is plastic and which isn't?

Is spandex sustainable? What about rayon? Will we remember that lyocell, for example, is superior to rayon or viscose when we're picking an item online or on a shop?

Where do I go to avoid polyester completely?

How do I remember which brand scores how much on what sustainability criteria?

Is all cotton good, even when we know that conventional cotton cultivation uses both pesticides and water heavily?

That's why moving from a society seemingly obsessed with fast fashion to one where slow fashion is the paradigm is the responsibility of the fashion brands, and of the policymakers and politicians.

There are several powerful levers for change for honest brands.

First, fast fashion should focus on its carbon emissions because it's a low-hanging fruit. The industry can accelerate and broaden plans to decarbonize cost-effectively and fast. According to McKinsey analysis, 80 per cent of the carbon emissions the industry needs to eliminate can be cut by initiatives that require top-down leadership, such as switching to renewable energy, and energy-efficiency and operational improvements. It's unclear what the industry is waiting for, given that at least 55 per cent of the measures would lead to net cost savings. Some larger global brands are showing the way, shepherded by the RE 100, a UK-based initiative (it stands for 100 per cent renewable energy), which organises them, provides them with the framework, then validates and showcases their progress. Medium-sized and smaller fashion and apparel brands around the world can follow their lead.

The RE 100 effort is an offspring of the Climate Group, an NGO that in 2011 decided to focus on changing the way people talked about climate change, from a narrative emphasizing disasters to one highlighting opportunities. In 2012, it launched its Clean Revolution Campaign, aiming for a dozen very large global companies to commit to a high-ambition pathway in order for business to positively influence the annual UN climate talks and to show what a difference leading corporations could make. It didn't work. The Climate Group struggled to get companies to commit €1 million to co-create a best-practice model: such businesses were much more used to funding and joining prescriptive initiatives that were pre-defined for them, and backing projects with clear deliverables rather than designing the deliverables themselves.

As the 2015 Paris climate talks inched closer, the Climate Group ditched its Clean Revolution Campaign and instead laid out what a company at the forefront of fighting climate change should be doing, such as being powered 100 per cent by renewable energy, doubling its energy productivity, and converting its fleets of trucks and automobiles to electric vehicles. In 2014, this inspired them to launch the RE 100 during Climate Week, an annual event they organised in New York City showcasing global climate action by governments and the private sector. IKEA and Swiss Re, the designated founding partners of RE 100, baulked just three months before launch at being alone. But with a lot of support from IKEA the weekend before Climate Week, the Climate Group managed to get a few more companies on board – including Mars and Unilever – and finally launched the initiative. Today RE 100 is the standard for large corporate climate action, and one of the world's most consequential NGOs. Over 300 companies with combined revenues of $5.6 trillion have made the commitment to 100 per cent renewable electricity, together already sourcing enough demand to power a medium-sized country and spending $100 billion to reach their goals.

Barriers to accessing renewable energy, such as regulations favouring fossil fuels and quasi-monopolies controlling an entire country's electricity market (as is the case in Indonesia and South Korea, for example), still exist in many markets, but are gradually being lifted thanks in part to pressure from RE 100 companies. Fashion brands that understood that being powered by renewable energy decreases costs include Gucci and the Estée Lauder Companies, both already powered 100 per cent by renewables; and Burberry and Chanel, on their way to 100 per cent by 2023 and 2025 respectively. Vast progress can be made quickly on decarbonization in fast fashion: Ralph Lauren joined RE 100 in

December 2019 and is cutting its fossil fuel use from 100 per cent to zero in just five years.

The United Nations is also using its convening power to push the fashion industry to decarbonize. Fashion, textile and clothing stakeholders worked during 2018 to identify ways in which to move forward on climate action, and issued the Fashion Industry Charter for Climate Action to guide themselves and others. Launched at the UN climate talks in Katowice in 2018, this charter was renewed at the Glasgow meeting in 2021. Its signatories – brands including Adidas, Burberry, Chanel, Decathlon, Gap, H&M, LVMH, Nike, Puma and others – are too few, however, and many are the same organizations that back RE 100. While the UN initiative is laudable, more needs to be done to include a broader universe of brands from around the world and to represent many more small and medium-sized enterprises.

At the same time as brands focus on their fossil fuel footprint and work to eliminate it, they need to become much more transparent. Given their dismal record so far, only country-level industry-wide legislation can do the job. This should include minimum design standards to ensure that garments that can be easily reused or recycled and a prohibition on burning unwanted clothes – a common technique fast fashion uses to protect their brands by faking scarcity. At the moment, it is practically impossible to recycle clothes because of the way plastic is combined with other fabrics inside them. We need to stop microplastics polluting rivers and seas by going to the source. Most importantly, the plastic paradigm must change: plastic simply cannot continue to be so cheap.

The fashion industry needs to change its business with a combination of positive and negative incentives. The materials used to produce clothes should be safe and come from renewable sources and their labelling should be transparent. At the same time, consumers need to make purchasing choices that are intelligent. Signal your displeasure by buying fewer textiles: the fashion industry, with its long track record of failing to enact any kind of self-regulation or reform, needs to be sent a much louder message, because in the end, the most efficient and effective way to root out bad corporate behaviour is through legislation. Meanwhile, buy clothes, but not too many. Preferably with natural fibres.

See also:
Chapter 1: Plastic Is Your New Diet

4

Your Cat Doesn't Need to Eat Fish

Periodically, mutilated dolphins wash up in their thousands on French beaches for all to see. Thousands more, uncounted, perish without ever reaching shore. That provokes indignation, then howling by locals and some media outlets. Then comes the silence. We forget about the lovely and super-intelligent descendants of land-dwelling mammals that first went to the sea 50 million years ago. We move on. This story then repeats itself every couple of years, in France, in Peru or elsewhere.

The dolphins' killers, however, don't take any breaks. They are shadowy people in massive boats that roam the oceans unregulated, armed with gigantic nets, indiscriminately catching every sea creature in their sight. They keep some for our consumption but kill millions and dump them back in the water. These are fish we never see except when they wash up dead on our beaches.

In the Indian Ocean, the killers might be pulling nets full of tuna, the primary catch they're after, mingled with discarded dolphins, giant clams, grey reef sharks and spiny dogfish. In the

Atlantic Ocean, they might have gone after lobster, mackerel or herring while sacrificing dolphins, sharks and small fish in vast quantities. Some of the finely meshed gill nets dragged by these industrial fishing boats can stretch 30 kilometres and haul out fish mountains of unfathomable scale.

Often the killers combine illegal fishing with human trafficking. They physically and psychologically abuse crew members, who are often poor migrants; in effect, modern slaves, desperate for work and forced to endure appalling conditions on board. The industrial fishing boats spend months or even years at sea, periodically encountering refrigerated mother ships to freshen up their supplies and offload their catch. They mostly get away with it because they specialize in hiding on the high seas – any area of the oceans not within 200 nautical miles (322 kilometres) of the coast of a country.

The high seas cover 43 per cent of our planet and make up 61 per cent of our oceans. A huge variety of creatures migrate daily from the seabed, sometimes hundreds of metres deep, to feed near the surface, then plunge back down to poop, contributing to a phenomenon known as the biological carbon pump, which captures vast quantities of carbon from the atmosphere to store in the depths. If we lost this biological carbon pump, or continued to weaken it, we would release enough greenhouse gases into the atmosphere to increase its CO_2 concentration by 50 per cent, accelerating climate change.

There are at least 60,000 industrial fishing ships, measuring up to 140 metres, trawling an area four times that covered by agriculture globally. No one actually knows how much of what they do is plain illegal fishing, if conducted in territorial waters, but it's not difficult to guess that most of it must be, or if it isn't, it should be. That's because the key techniques industrial fishing relies on are inherently murky and environmentally destructive, and no navy has the capability to monitor, let alone police, the high seas.

Industrial fishing boats use two principal techniques to fish: bottom trawling and longline fishing. Bottom trawling involves dragging giant nets across the ocean floor, catching fish randomly and indiscriminately while destroying ecosystems along the way. Even worse, sometimes two vessels drag nets stretched between them to maximize the catch, which increases the destructive power. When longline fishing, boats drag lines up to 15 metres long with multiple hooks on them and similarly randomly catch fish and whatever else is grabbed along the way. In both techniques, an enormous amount of fish and other catch are thrown away because they're not what the boats were after. The result is the deliberate destruction of vast quantities of fish and other sea creatures.

Many people have been trying to get to the bottom of the illegal fishing trade. For almost two decades, NGOs like Greenpeace and Oceana, for example, have launched investigations, initiatives and public relations stunts. Independent filmmakers have produced documentaries, such as *The End of the Line* by Rupert Murray, and *Seaspiracy* by Ali Tabrizi. Sporadic access to satellite data has helped as well. An NGO collaboration examined 22 billion hourly signals from the automatic identification systems of 70,000 industrial fishing vessels to show that just five countries – China, Spain, Taiwan, Japan and South Korea – were responsible for 85 per cent of industrial fishing. But although that scrutiny shines a light on the secretive and shadowy nature of the industry, its criminal mindset and the desperate need for it to be regulated or shut down, it doesn't seem to affect it or stop it all.

The result is overfishing powered by an absurd $35 billion in annual subsidies – 20 per cent of which is for cut-price fuel – from the US, the European Union, China, South Korea and Japan. Without subsidies, 54 per cent of all the fishing on the high seas would be unprofitable, because it would be too expensive in terms of the required fuel to get there and back to port.

Squid fishing on the high seas, for example, would instantly stop without these subsidies because the sale price of squid doesn't cover the cost of the fuel required for a long-distance round trip. China dominates squid fishing and catches at least 50 per cent of all the squid from the high seas, but their government spends billions annually to subside their diesel, allowing them to travel further and to build more advanced fishing boats. In addition, China mobilizes its satellites and naval intelligence – at no cost – to help its fishing fleets find the most abundant squid resources.

The scale of the destruction is massive. Today, at least 37 per cent and perhaps up to 82 per cent of the 1,300 commercial species of fish are being depleted faster than they can reproduce. In addition, the industry kills 50 million sharks every year while bottom trawling or longline fishing – at zero cost to ship owners and operators. Environmental destruction is free of charge.

We have solutions and they don't involve stopping eating fish, or ceasing to fish. The world is perfectly capable of sustainably managing its fish resources, and it must. The fishing industry has been employing people since ancient times. Today, some 250 million people are directly involved in it. One billion people, mostly in developing countries, derive their primary protein from fish, while 3.3 billion receive 20 per cent of their animal-protein needs from seafood. Campaigning for people to stop eating fish is tantamount to trying to deprive some of the world's poorest of their essential sources of protein and income.

In addition, many coastal populations – in Indonesia, West Africa and the Pacific islands, for example – live in harmony with the seas and are already battling climate change impacts. According to the UN climate science panel, the Intergovernmental Panel on Climate Change, the fishing-dependent people of the Pacific islands are experiencing increased droughts, coastal flooding and erosion, a decrease in the availability of drinking water and changes in their rainfall patterns negatively affecting their food production.

What we need is for oceans to be managed in a sustainable way. To get there, the most important action we can take is at a government level, by agreeing to turn at least 30 per cent of our oceans (and ideally 50 per cent) into marine reserves; in other words, areas protected from fishing and other human incursions, such as deep seabed mining in search of new sources of minerals. Today we protect less than 5 per cent of our oceans. The 30 per cent threshold would keep a substantial portion healthy and restore fish populations. This would work in a similar way to how we should be fighting deforestation: prepare the ground and leave trees, or fish and other sea creatures, to do their thing.

In addition to protecting the high seas, we need to apply a scientific approach to existing fisheries within national boundaries. This would involve setting strict – and enforced – catch limits, while closing key habitat areas to support the ocean's ecosystem. It would help oceans begin to recover by reducing overfishing and limiting the senseless fishing, then dumping, of by-catches.

The existing system is broken. It is built around multiple regional fisheries management organizations set up in and around countries that have fishing interests in a specific area. Their record is one of failure, as documented by the rise of industrial fishing around the world, even though some have the power to set rules; for example, regarding fishing limits.

On the face of it, protecting 30 per cent of our oceans shouldn't be too hard: just 30 countries and the European Union govern 90 per cent of our global fish catch. But we're swimming in a maze of initiatives, none leading us forward fast and every one partially dependent or waiting on another. Efforts are under way, for example, to protect some of our oceans via the ambitious Intergovernmental Conference on the Protection of Biodiversity Beyond National Jurisdiction, convened under UN auspices to, as it says on the tin, protect biodiversity specifically via an international legally binding treaty to protect the high seas and their marine life. In theory, this is intended to complement another UN instrument, the Convention on Biological Diversity (designed to protect wildlife), which only applies within a country's border or to vessels carrying its flag.

The UN Convention on Biological Diversity is at its fifteenth annual meeting. The Intergovernmental Conference seeking an international treaty to protect marine biological diversity in the high seas is at its fourth annual meeting. It took more than 21 annual meetings to agree the 2015 Paris Agreement on climate change, setting out a global framework to limit global warming to below 2° Celsius. We had waited for governments to act on runaway climate change since 1992, and at last they delivered something. The result, while representing solid progress on what prevailed before the agreement went into force, wasn't binding or enforceable. Furthermore, the United Nations periodically reviews progress towards the goals agreed in Paris, and in one report it said that 191 countries are still sleepwalking into a deeper climate catastrophe because their collective plans deliver emissions 16 per cent higher in 2030 compared to 2010, as well as an expected 2.7° Celsius of warming

The oceans need us to do better than we have done on climate change. We must make the Intergovernmental Conference on the Protection of Biodiversity Beyond National Jurisdiction both binding and enforceable, because at a stroke we would have a mechanism to aggregate enough marine reserves to fight back against industrial fishing, and in time wind it down substantially. Similarly, the UN Convention on Biological Diversity needs to stop the Groundhog Day routine of annual meeting followed by annual meeting without a result. It's already enough that yet another two UN initiatives related to the oceans, the High-Level UN Ocean Conference and the not yet properly named global convention on plastic litter, don't look like achieving anything concrete any time soon.

To add to the confusion, there is also the High Level Panel for a Sustainable Ocean Economy, made up of the heads of states and governments of 14 countries who say they want to put sustainability at the heart of ocean management, production and protection; the annual One Planet Summit, which is sometimes focused on biodiversity, including the ocean; and the High Ambition Coalition for Nature and People, made up of more than 45 countries, apparently there to champion a global deal to halt the growing loss of species and vital ecosystems on land and sea.

One of the most effective short-term measures needed is to end subsidies for fishing in the high seas. Negotiations have been ongoing for two decades, and it is unconscionable that pressure is put on people to stop eating fish instead of on governments and the World Trade Organization (WTO) to do their job. Indeed, the WTO repeatedly failed to stay on course to comply with its mandate to eliminate by 2020 fisheries subsidies that contributed to overfishing, overcapacity and illegal, unreported and unregulated fishing.

A combination of marine reserves covering large parts of the high seas, the removal of fuel subsidies for long-distance fishing and a science-based approach to fishing within national boundaries would result in an increase in global fish stocks, making wild fish more accessible to artisanal fishermen and hungry people around the world while also providing protection for important ocean ecosystems, habitats and biodiversity.

More and better labels on fish would also help, in order to ascertain origin and traceability. Consumers could then favour fish that come from smaller boats using fishing lines, rather than nets or longlines. Meanwhile, those who want others to stop eating fish should focus instead on the fact that Australian cats, for example, eat 13.7 kilos of fish each year, more than the 11 kilos of fish and seafood eaten on average by human beings.

See also:
Chapter 26: The Royal Baby Versus Biodiversity

5

Your Fresh Air Is Asphyxiating You

Two 2020 scientific studies should be front-page news every day. According to the first one, air pollution is linked to 11 per cent more deaths from COVID-19, while the second one said that 15 per cent of worldwide COVID-19 deaths may have resulted from dirty air and the damage it causes to our heart and our lungs. This latter study also estimated that 27 per cent of coronavirus deaths in China are attributable to air pollution, 26 per cent in Germany, 18 per cent in the US and 14 per cent in the UK. In other words, air pollution makes COVID-19 worse by increasing the number and severity of cases.

Intuitively, this makes sense, right? If you're breathing dirty air, it can only make the impact of a respiratory illness from a virus worse.

Well, we are almost invariably breathing dirty air: fresh air is a myth. We think we encounter fresh air, on a personal level, if we walk out to a park or a green space in a city, for example, except that is absolutely not what happens: the air is heavily polluted, and worse, we've known that for decades and aren't doing much about it.

Many years before Greta Thunberg, there was Severn Cullis-Suzuki. Severn founded an environmental children's organization in 1989, aged nine. Three years later, she attended the Earth Summit in Rio de Janeiro, where she became famous as the girl who silenced the world for five minutes when she said in her speech that she was afraid to go out in the sun because of the hole in the ozone layer; that she was afraid to breathe the air because she didn't know what chemicals were in it. Three decades later, she should still be afraid.

Both studies linking air pollution with the incidence of coronavirus deaths are peer-reviewed and have been published in academic journals – *Science Advances* and *Cardiovascular Research*. One focused on the United States, while the other had a global scope. Both confirm that Severn, correct in 1992, is still correct today. We too should be afraid to breathe the air, because we most definitely don't know what chemicals are in it. That's because we haven't done much about it since.

We know now that there are three types of air pollutants that cause terrible health problems. The first is called particulate matter (PM). There's a small menu of PMs, but PM 2.5 and PM 10 are of particular interest. The numbers refer to their tiny size, which is measured in micrometres. A single hair is 70 micrometres in diameter, so we're talking about pollutants one seventh to one thirtieth smaller than that and which we can't see with the naked eye. Particulate matter worsens heart and lung disease. It contains microscopic solid or liquid droplets that are so small as to be inhaled by humans without us noticing. They go deep into our lungs and some seep into our bloodstream. The second type of air pollutant is nitrogen dioxide. This causes a flare-up of asthma or its symptoms – for example, coughing and difficulty breathing. The third type of air pollutant is ground-level ozone. This forms when heat and sunlight allow a reaction between two other pollutants, nitrogen oxide and volatile organic compounds, gases emitted from some solids or liquids.

You're probably not going to be surprised when I tell you what all three variants of air pollution have in common: each is a by-product of power plants burning coal, gas and oil; petrol or diesel cars, buses and trucks; central heating systems in homes and offices if they are gas-powered; and industrial plants powered by fossil fuels.

At its core, most of our air pollution is generated by burning oil, gas and coal. There are many other sources of that pollution, such as dust from construction sites, unpaved roads and wildfires; however, these play a minor role. The fundamental driver of air pollution is the burning of fossil fuels over the past 150 years, using the air as a free garbage can. Because that pollution is invisible to the naked eye, we are fooled, when looking out of our windows, into thinking we can go out and walk around in a city park to breath fresh air. In reality, we are inhaling a lethal cocktail of PM 2.5, PM 10, nitrogen dioxide and ground-level ozone.

Air pollution isn't new. Ancient Romans built outstanding water piping infrastructure in their cities, but to do so, they had to mass-produce lead, a soft, malleable metal they called *plumbum* (the word is the origin of 'plumbing'). Lead, however, is extremely toxic to humans, and the Romans managed to increase its concentration in the atmosphere by 10 times, heavily polluting their tap water in the process. For comparison, we've upped that level of lead pollution 10 times during the period we were using leaded petrol.

Alarmingly, according to the World Health Organization, nine out of ten citizens worldwide today are breathing dirty air, and the poorer the region or country, the dirtier the air. This translates into more deaths from strokes, heart disease, chronic obstructive pulmonary disease, lung cancer and acute respiratory infections. Air pollution from the burning of fossil fuels is responsible for one out of six deaths each year, or approximately nine million people (about 55 million people die each year worldwide, pandemics not included).

And that's just the deaths. Billions of other people are adversely affected: asthma sufferers, people with weak lungs, weak hearts and generally weak health. Nobody producing or promoting fossil fuels is paying for this assault on the well-being of the population at large. What is particularly frustrating about the myth of fresh air is that we lose out on the actual benefits cleaner air would bring: less traffic, more habitable cities and towns and fewer days off sick. In addition, the resilience of our healthcare systems would be significantly enhanced because of the enormous amount of time and money spent on dealing with illnesses related to pollution from fossil fuels.

New Delhi, for example, routinely declares public health emergencies, shuts down schools and distributes millions of protective masks, not because of any virus, but because of dirty air. The pollution from fossil fuels is terrible all year round, but the city's government only acts when it becomes clear for all to see with their own eyes, as the haze from seasonal wildfires and crop burning combines with pollution from cars, buses, scooters and industry to create air that is literally poisonous.

In China, air pollution rather than climate change was the major factor behind the country's war to clean up its environment. The Chinese population in cities saw the problem with their own eyes when pollution from industrial activities combined with that from transport and seasonal weather. They then demonstrated en masse: in the 2010s, China routinely clocked 80,000 environmental riots a year. Nothing less than the legitimacy of the communist party was at stake, so the government began to restrict the number of high-polluting cars and buses on the roads, heavily pushing electric buses; as well as periodically shutting down polluting industries such as steel. China also turned several hundred million petrol-fuelled two-wheelers into electric scooters. It understood the direction of travel: namely to stop burning coal and other hydrocarbons.

The correlation between China's prodigious consumption of fossil fuels and its high death toll from air pollution was clear. While air pollution kills one in ten people in the West, that ratio drops to one in three in China and India. It's not a coincidence that the world's most congested cities are those with the dirtiest air: all of the top 15 most polluted cities in the world (from a PM 2.5 perspective) are in China and India.

Today, it's no longer an excuse that the pollution particles are so small as to be invisible, or that science hasn't caught up with the impact of pollution from fossil fuels. At the same time, we've been putting pressure for decades on wild animals as we take over where they live and destroy their ecosystems. One of the most dramatic results has been that many more animal pathogens have jumped from their world to ours: Ebola, anthrax, bubonic plague and others. Coronavirus is just the most recent example of nature trying to tell us that enough is enough. Yet we've even managed to make pandemics worse by eliminating fresh air from the reach of 9 out of 10 humans on the planet.

We need to take much more forceful action. At an individual level, we can walk more, cycle more and switch to electrifying everything we can, from our cars to our boilers. But more importantly, as a collective, we can vote for fresh air by fighting climate change, because both require us to cut by 90 per cent the oil, gas and coal that we use today.

In countries, regions, cities and towns where it can be done, push for anti-pollution measures to be added to the ballot paper, to empower citizens to vote for plans to end illegal air pollution; to vote for diesel to disappear; to vote for changes to road taxes where these exist, so that anybody driving a diesel or petrol car is pushed to stop; to vote for more pedestrian zones; to vote for more bicycles, and to vote for more infrastructure to support electric cars, buses and scooters, especially in poorer countries, where scooters represent the main mode of transportation. In Jakarta, Ho Chi Minh and Mumbai, for example, all petrol scooters should be replaced by electric scooters.

If you can't vote, join a peaceful protest.

If the democratic mandate is there, the impact on fresh air and health services around the world will be gigantic.

6

We Don't Have Time to Overthrow Capitalism

On 18 June 1999, an anti-capitalism protest in London, timed together with dozens of other protests around the world to coincide with the start of a summit of the club of richest nations, referred to then as the G8, turned violent. Demonstrators attacked buildings and temporarily occupied the London International Financial Futures Exchange. Dozens were arrested and over 100 were injured. Today, protests of this kind still try to capture the zeitgeist of global system failures such as climate change and biodiversity loss. They shouldn't. Only capitalism is likely to provide the answers to the climate emergency.

Individuals and private businesses should privately own land and the means of production, alongside government-owned properties and capital goods. Markets and price signals should direct where the money goes, in line with supply and demand. With very few exceptions, every country in the world practises capitalism in one form or another, with varying degrees of government regulation of business and safety nets for citizens, many of which are checked periodically at the ballot box. Even Cuba would be hard pressed to argue that it's not beginning to adopt capitalism, by opening the overwhelming majority of economic activities in the country to the private sector.

Capitalism is flexible and adaptable to the will of the people in democracies or to the will of autocrats in authoritarian regimes. Social safety nets can be broad (as they are in Scandinavia, for example) or narrow (as they are in the United States). Excesses can be tamed (as we do periodically when the pendulum swings from laissez-faire to more robust government regulations). Financial crises can come and go. Problems with the system can be addressed.

Nonetheless, pundits and commentators regularly shout that what we need to do to solve climate change is nothing less than overthrow capitalism. These are often armchair pontifications made from the comfort of Western homes. A typical line might be labelling capitalism as an imperialist system of colonization of the planet, or casually concluding that it is incompatible with the survival of life on earth. Habitually these accusations are made without providing an alternative system – especially one that can be implemented globally, with popular support, within a decade and simultaneously tackle our key systemic failures of climate change and biodiversity loss.

We don't have time to switch almost eight billion people to plant-based diets, or to stop everyone flying. We don't have time to overthrow capitalism either, even if that was the right thing to do, which it isn't. Do people seriously think that the US, Britain, India or China, for example, are going to get rid of capitalism any time soon? Calls to do so are counterproductive. They divert attention and resources from solving the climate change and biodiversity crises. They also promote internecine strife within a global climate movement that doesn't have room for such conflict.

What is true – and welcome – is that capitalism has been evolving, as concerns over environmental, social and governance criteria (ESG) have intensified. In a challenge to Milton Friedman's argument that the sole purpose of a corporation is to maximize shareholder value, the US Business Roundtable, representing 181 chief executive officers, redefined in 2019 the purpose of the corporation. It committed to new principles, including protecting the environment by embracing sustainable practices across and within businesses and their supply chains.

The move was a response to pressure from consumers as well as staff, as a younger generation, which sees climate change as an existential threat, wields greater power. Commitment to ESG is sometimes real, sometimes no more than a fig leaf – greenwashing – but most companies in the Western world today purport to adhere to a more responsible capitalism. That's already a step-change, compared to the previous 300 years.

Instead of naïvely calling for the abolition of capitalism, we should focus on holding companies to their commitments and pushing more towards sustainability, whether in the production of goods or the supply. We should also focus on reducing the power of lobbies to derail climate action. Capitalism is perfectly suited to regulate the system from within. For example, the largest five stock-market-listed oil and gas companies spent $1 billion over the past three years lobbying to delay, control or block policies to tackle climate change. This includes funding climate-sceptic think tanks, fake experts, bots and distorted research in order to buy time to sell more oil and gas, as well as producing misleading advertising. This doesn't include the lobbying budgets of the state-controlled oil companies. Saudi Aramco made $110 billion in profits in 2019 and we have no idea how much of it was spent on influencing policies, the public and politicians around the world. This lobbying could be made illegal at the stroke of a pen if company directors were held personally liable for knowingly perpetrating harmful climate and biodiversity impacts.

We also need solutions that can be delivered swiftly and decisively, rather than counterproductive proposals that are a fantasy. First among these is abolishing fossil fuel subsidies. The Intergovernmental Panel on Climate Change found that emissions from fossil fuels are the overwhelming cause of climate change, accounting for 89 per cent of global CO_2 emissions. Yet according to the International Monetary Fund (IMF), we continue to this day to subsidize the production and burning of coal, oil and gas to the extent of $5.9 trillion, or 6.8 per cent of global gross domestic product. That's $11 million every minute of every day paid to Big Oil to make the climate emergency more acute. Worse still, this is expected to rise to $6.4 trillion in 2025. These sums aren't even properly labelled when they are called subsidies; what they really are is a contribution of our cash, our lungs, our planet and our lives to the oil, gas and coal industries, for free.

Everyone is at it. Not one of the 191 countries party to the 2015 Paris Agreement prices its fuels sufficiently to reflect their true costs, including their environmental destruction costs. Explicit subsidies – in cash – account for $450 billion of the $5.9 trillion, with the balance being implicit subsidies (for example, underpricing for local air pollution and climate damage). Cash subsidies are projected by the IMF to increase to $600 billion by 2025, a fact almost beyond belief in the midst of a climate emergency.

Cutting fossil fuel subsidies would single-handedly cut global CO_2 emissions by 36 per cent and prevent at least 2.2 million deaths a year from dirty air, a 24 per cent reduction from the 8 million premature deaths a year we are experiencing today because of fossil fuels, while raising substantial revenues worth 3.8 per cent of global domestic product.

Wouldn't activists' time be much better spent fighting fossil fuel subsidies rather than going after something that won't happen? Every minute counts in the fight against climate change, and dismantling the fossil fuel edifice is an achievable and realistic goal in the time we have left to limit warming

The climate movement is already very broad and includes hundreds of thousands – perhaps more – of well-meaning and competent people. It's also unwieldy and unfocused. Climate advocates today come in numerous varieties. Advocacy organizations include 350.org, Greenpeace, Friends of the Earth, the World Wildlife Fund, the Sierra Club, Extinction Rebellion and many others. Some of these are enormous. Greenpeace alone has more than three million donors and annual budgets of over $100 million. Some are organized in global networks (such as the Climate Action Network, a worldwide network of over 850 NGOs in more than 90 countries). Then there are large global conservation-focused organizations such as the Nature Conservancy, endowed with $8 billion in assets, and Conservation International, whose annual budgets are $150 million. All are supported by the vast intellectual power of a maze of think tanks publishing thoughtful, fact-based research about our climate crisis, including Tsinghua University (China), the Energy and Resources Institute (India), the Institute for Sustainable Development and International Relations (Europe), the World Resources Institute and Brookings (US) and many, many others.

On university campuses around the world, a 'divest' movement (calling for divestment from fossil fuel investments) is pushing as hard as it can. In parallel, shareholder activists such as CERES, Share Action, the Carbon Disclosure Project (CDP), the Global Investor Coalition on Climate Change and others are lobbying for sustainable business and investment practices, increased disclosure of climate risks, and government policies and investment practices that address the risks and opportunities of climate change. Then there are the new kids on the block: net-zero-focused organizations such as the Science Based Targets initiative (a partnership between CDP, the United Nations Global Compact, the World Resources Institute and the World Wide Fund for Nature) and the We Mean Business Coalition. An extensive and costly UN infrastructure is also fully deployed in a spider web of supranational bodies, international conventions, accords, protocols, clubs and investment funds, in addition to the World Bank and the IMF, who are increasingly – and rightly – fully signed-up members of the global climate movement.

Notwithstanding the efforts deployed by this large climate movement, emissions are still rising – although there is reason to hope that we will bend the curve soon. We must stay focused on delivering on the mission of these multitudes of organizations, backed by tens of millions of citizens worldwide, especially because change on a global scale is being delivered. There is no global precedent to the changes under way, with China, Japan, South Korea, the US, the European Union, the UK and others all committing to reduce their emissions to zero within a few decades.

It is far better to concentrate our collective efforts on levers of change such as abolishing fossil fuel subsidies, and to seek to more rapidly implement sustainable ways forward for our transportation sector, for industry, for energy, for water and for food. We need to stop being a wasteful throwaway society. We need to build circular economies. We are, in fact, doing just that through a combination of fundamental shifts at government level as well as changes to the price of money going to oil, gas and coal investments and products. We can't afford to waste time and resources on ideological battles straight out of the nineteenth century. Far better to continue to improve what we already have.

7

Hydrogen Makes Up
70 Per Cent of the Universe;
I Didn't Know That Either

Hydrogen is the most abundant chemical substance in the universe. It's also one of the most difficult to conceptualize. You can't see it. It has no colour, no smell and no taste. It's not toxic. It's not a metal, and it burns very easily in its pure form. Yet it's in all plants and living things in general; we're approximately 63 per cent hydrogen.

On earth, hydrogen is usually a gas, and it's one of the parts that make up a water molecule, the H in H_2O. It is also the fuel that powers the sun and other stars. In addition, it's an extremely versatile element that can be used to create complex hydrocarbons like natural gas and oil.

As the imperative to reduce man-made emissions started to gain more momentum after the Paris Agreement in 2015 with the goal of limiting the rise in global warming to well below 2° Celsius compared to pre-industrial levels, the world started to talk a lot more about hydrogen because of its ability to help in decarbonizing the world and ending our addiction to fossil fuels.

This isn't the first time we've pinned our hopes for a cleaner future on hydrogen. As early as the 1970s, it was presented as a key plank in decarbonizing the world as a potential replacement for fossil fuels. Japan picked up the mantle after the oil shock of 1973 (when Arab oil-exporting countries embargoed exports to the United States and the resulting increase in oil prices jolted the Japanese economy) and investigated its usefulness as a backup energy source, but with no useful results.

Hydrogen then acquired magic potion status when *The Hydrogen Economy*, a book by Jeremy Rifkin, was published in 2002, though once again without tangible results in the real world: you were hard pressed to find anything hydrogen-powered anywhere. Instead, hydrogen evolved as a feedstock for industrial and chemical processes, and that's where most of it is used today; for example in refining, in petroleum products, and in the production of methanol and that of ammonia for fertilizers.

Today's excitement comes from the idea of using hydrogen as an energy storage medium. It is rich in energy and can, for example, be used to power cars and trucks. It can be converted into electricity using a fuel cell or burned like natural gas to produce power and/or heat. The by-product of both processes is harmless water vapour. But hydrogen can also be kept indefinitely, so it's a potential long-term form of energy storage.

That's why it is now being touted as the low-carbon solution for sectors that are more difficult to wean off fossil fuels because of the costs and the technological challenges involved; for example, steel production, shipping and long-haul trucking.

However, the strangest thing about hydrogen is that despite the fact that it is the most plentiful element in the universe, we have no natural, pure supplies of it. If we want more hydrogen because we want to use it as energy storage or in an industrial process, we have to make it. Therein lies our challenge.

There are three principal ways to make hydrogen, and these result in either green hydrogen, black (sometimes referred to as grey) hydrogen or blue hydrogen.

Green hydrogen is made from renewable energy. That's good, because it means it releases close to zero climate-warming emissions, and we can use it in place of fossil fuels – for example, in the manufacture of steel. Steel accounts for approximately 9 per cent of global emissions (more than the emissions of all the cars in the world), because we use it everywhere, including in automobiles, buildings, everyday tools and appliances, ships, trains and weaponry. Like cement, steel is one of those materials you don't really notice is all around you until you start to look for it. Then you'll find it's just about everywhere: It's on your dinner table in the form of cutlery, in the wind turbines on the hill, in the bridges you cross every day, in the buildings you enter, in the cars you drive, in the food cans you buy.

Steel requires a lot of energy to make. On average, every ton of steel demands about two tons of carbon dioxide. It needs high temperatures and intensive energy consumption to break the chemical bond in iron ore to get pure iron. For 2,000 years, we've been using coal to break that bond, a reaction that requires a reagent to bind with oxygen. This generates CO_2 as a by-product. But current technology allows green hydrogen to remove the oxygen from the iron oxide. This cuts carbon emissions by 95 per cent compared to how we've been producing steel since the Industrial Revolution. BMW, for example, is already buying green steel from a Swedish start-up to cut carbon emissions in its steel supply chain.

Shipping is another industry poised to be decarbonized using green hydrogen. The International Renewable Energy Agency (IRENA) has predicted that green hydrogen-powered e-ammonia – a 100 per cent renewable and carbon-free means of transporting hydrogen, which can then be used as a green fuel – will supply the backbone of the decarbonization of the shipping sector. This 'powerfuel' would contribute 60 per cent of shipping's mid-century decarbonization effort under IRENA's pathway to a world that will see global heating capped at 1.5° Celsius.

Black (sometimes referred to as grey) hydrogen is derived from fossil fuels. That's bad, because it represents new and potentially rapidly rising ways to continue to use oil, gas and coal and thus produce more greenhouse gas emissions. Blue hydrogen is also made from fossil fuels, but with the added promise that it will be combined with fantastical carbon capture and storage or tree planting to reduce its climate impacts. That's also bad, because carbon capture doesn't work very well, and as we discuss later, in Chapter 11, tree planting is typically a fig leaf used as an excuse to continue to pollute. Arguably, blue hydrogen is therefore worse than black hydrogen because it's trying to greenwash untruths in plain sight.

When looking at hydrogen as a magic potion to solve global warming challenges, it's important to keep things in perspective. The global hydrogen market is worth some $130 billion today, most of which is black hydrogen; this is about the same size as the global pizza market. In addition, the green hydrogen we desperately require, because it has zero emissions, hardly exists.

Currently, 99 per cent of the hydrogen we produce is derived from dirty fossil fuels, releasing more climate-warming emissions than the UK, France and Belgium combined.

We need to turn green hydrogen into a trillion-dollar market furiously fast, without getting distracted by either black or blue hydrogen – a distraction that is bound to make its presence felt because of the power of its backers, producers of oil and gas. Research from Goldman Sachs and Bank of America, for example, indicates that in a near-zero-emissions world, green hydrogen probably needs to become a $2 trillion per year market by 2050.

That's why, after a hiatus dating back to the early 2000s, hydrogen is back in vogue and taking up a large space in the conversation about fighting climate change. Unlike the early 2000s, however, the conversation is also turning into action, with governments and corporations announcing investment initiatives and projects by the bucketload.

Japan is trying to build a hydrogen society by 2050, aiming to transform its entire economy by harnessing hydrogen for pretty much everything. The country already has 169 hydrogen refuelling stations, more than any other, to power hydrogen fuel cell cars and trucks, those these haven't turned up at scale yet.

The European Union, the world's third largest polluter after China and the United States, introduced its EU Green Deal in 2019, possibly the world's most ambitious plan to decarbonize completely by 2050. It contains a roadmap of more than 50 actions that Europe is taking to get there, including an ambitious green hydrogen strategy. The 2022 Ukraine war has provided further urgency to accelerate Europe's green hydrogen strategy, a key plank in breaking the continent's reliance on Russian gas.

China, which wants to be carbon neutral by 2060, has also announced that it will make its steel industry – the largest in the world and responsible for 15 per cent of the country's climate-warming gases – green with the help of hydrogen.

In Australia, the Fortescue Metals Group is embarking on the world's largest project to produce green hydrogen as well as ammonia from 235 gigawatts of yet-to-be-built wind and solar power projects. That's about five times Africa's entire renewable energy capacity, or all of America's renewables.

In total, several hundred projects have been announced around the world, with proposed investments of $300 billion just this decade. In addition, more than 30 countries are implementing national hydrogen strategies. However, it's crucial that the hydrogen is produced sustainably, and that's where the danger lies: not all hydrogen announcements are equal, and any black or blue hydrogen project or investment promotes the continued use of oil, gas and coal while giving the impression of acting against climate change.

Japan is a case in point. While it states that it wants to power multiple economic sectors with hydrogen, it is setting about it in a way that will make efforts to fight climate change more difficult. A key plank of its strategy is using Australian brown coal (the dirtiest kind of coal) and Australian coal-fired power plants to produce hydrogen, which would then be liquefied and shipped to Japan on supertankers. The result is the perpetuation of a coal industry we need to retire, while launching brand-new and purpose-built supertankers fuelled by other fossil fuels.

By contrast, one of the world's first green hydrogen production plants is being built – at commercial scale – in the port of Ostend in Belgium, powered only by surplus offshore wind energy. That's good green hydrogen. Impressively, Belgium was number four in Europe in offshore wind capacity in 2021, behind the UK, Germany and the Netherlands, even though it is endowed with a coastline measuring only 65 kilometres, and it is planning to at least double its offshore wind capacity by 2030 to facilitate the ability of its chemical and other heavy industries to decarbonize. These offshore wind farms generate excess renewable energy that is not absorbed by the grid. That's because they produce a lot of energy at night, when we're sleeping with our lights off and therefore not using much of it. All that excess power is often curtailed and wasted rather than stored for future use. But in the case of the Belgian hydrogen project, it is diverted into the production of clean green hydrogen.

Another successful case study can be found on Scotland's gorgeous and remote Orkney islands, which produce more clean energy than the locals need. Orkney is endowed with rich wind resources and is the birthplace of the UK wind energy industry: the country's first grid-connected wind turbine was installed there in 1951. The islands are using their surplus clean energy from wind and tidal power to generate green hydrogen. They are on their way to having the world's first seagoing car and passenger ferry fuelled only by hydrogen, and they already have a hydrogen refuelling station for hydrogen fuel cell vans.

Over in Luleå, a city in Sweden's Lapland dating back to the seventeenth century, and the capital of the northernmost county in the country, green steel plants are being developed by Swedish steelmakers. Clean hydro power is plentiful and cheap, close to the country's Kiruna mine, one of the largest and most modern underground iron ore mines in the world and a source of raw material for Swedish steel.

Ostend, the Orkney islands and Luleå are the exceptions, though, because green hydrogen is still expensive. As renewable energy gets cheaper everywhere around the world, green hydrogen will also get cheaper and can increasingly serve as an energy source for electricity, transport, heat and fuel and as a raw material for industrial purposes. But while costs are dropping and are expected to continue to do so (specialists think green steel will be competitive with steel manufactured using fossil fuels by 2030), we're not there yet.

The process used to create hydrogen is called electrolysis, which refers to the use of electricity to split water molecules into hydrogen and oxygen in order to isolate the hydrogen before it is liquefied and transported. It can then be burned at a later date to generate electricity. Unlike petrol or diesel, burning hydrogen does not produce harmful by-products.

In a truck, for example, hydrogen combines with oxygen to produce an electrical reaction that powers the engine. Water vapour only is emitted from the exhaust, with no harmful emissions whatsoever, and therefore no air pollution or greenhouse gas emissions that contribute to climate change. In long-haul transport, hydrogen fuel cells can combine with electric batteries to cover truly long distances, much longer than the typical range of leading-edge batteries. Hydrogen can also be used – at least hypothetically – to heat buildings, power electrical facilities and propel trains, ferries and cargo ships. Another benefit of hydrogen in a decarbonization strategy is that once you produce it, you can store it and move it around to where it's needed on a large scale.

Obfuscation by fossil fuel interests is de rigueur, and whenever hydrogen is encountered, we need to be suspicious. A few general rules may be helpful.

Rule number one is that often when hydrogen is suggested, it's probably an oil and gas company, or an industry association grouping several of them, peddling its product in an indirect way. Take the Oil and Gas Climate Initiative (OGCI), for example. Its members are the usual suspects, all the big oil and gas companies including ExxonMobil, Shell, BP, TotalEnergies, Chevron, Occidental Petroleum, China's CNPC, Italy's Eni, Norway's Equinor, Mexico's Pemex, Brazil's Petrobras, Spain's Repsol and Saudi Aramco. While their declared strategy is working towards net zero by funding companies that look like they're contributing to the fight against climate change, they are in fact developing innovative ways to sell more oil and gas.

One of the companies backed by the OGCI is Achates Power, a developer of apparently radically improved internal combustion engines that increase fuel efficiency, reduce greenhouse gases and cost less than conventional engines. That may all be true, but they're still petrol-powered internal combustion engines: serious rearguard action to protect oil and gas exploration and production disguised as progress.

Another example of a company backed by OGCI is Clark Valve, a developer of valves that fight fugitive methane emissions. It's encouraging that the problem of fugitive methane emissions is being tackled, but the entire premise of this investment is once again to preserve the status quo, by ensuring oil companies can position themselves as producers of clean natural gas when even the International Energy Agency, traditionally a bastion of fossil fuel interests, now says we can't afford more.

If your strategy is rearguard action and obfuscation, black and blue hydrogen are prime real estate: hydrogen is attractive because it's zero carbon, but it can be produced from fossil fuels, and this ensures business as usual for oil and gas. Naturally, the oil and gas industry is seeking to co-opt the global hydrogen industry by rushing to produce black and blue hydrogen. A case in point: the black and blue hydrogen lobby spent almost €60 million trying to influence Brussels policymakers in 2020, meeting with key commissioners and their cabinets more than 13 times a month between December 2019 and September 2020. That's why references to hydrogen as a critical tool in decarbonizing entire sectors should be met with suspicion until it is established that green hydrogen is the fuel being produced. The coal industry has employed a similar technique to postpone its inevitable demise by several decades: it strategically used the idea that you can still burn coal, but you can capture its harmful greenhouse gases through carbon capture and storage technology that's just around the corner, even though it isn't.

Rule number two is that if done properly, hydrogen is safe. A standard worry is that it is very flammable and yields explosive mixtures with air and oxygen. That worry doesn't withstand scrutiny. Hydrogen is safer than fossil fuels – gasoline, for example – because it requires higher concentrations of oxygen to make it explode. In addition, we've been using it as a fuel for over 50 years, more than enough time to develop robust safety standards; it is 'the signature fuel of the American space program', according to NASA, and has been propelling rockets since the 1950s.

Rule number three is that green hydrogen is always a very good idea and will inevitably prevail – versus the other colours – on climate action grounds. Although it forms a tiny sliver of the global hydrogen market today, the continued drop in the cost of solar and wind power means it is becoming more competitive by the day. Initially green hydrogen will most likely be used selectively – for example, to replace diesel in long-haul trucking – until much more is produced at scale from renewable energy.

For now, hydrogen is not for everyone and not for everywhere. It's an option that we can use to help us decarbonize really difficult bits of the energy system, but only once we have enough cost-competitive green hydrogen at scale. We will use it as a chemical and energy agent in steelmaking; in the manufacture of fertilizers; and in multiple transport applications including long-haul trucking, aviation and shipping. Meanwhile, we must avoid building new infrastructure that's dependent on hydrogen produced from fossil fuels, whatever its colour palette. We need to be careful to call out the greenwashing of black and blue hydrogen, which risks misdirecting the support of governments around the world.

We can meet all our energy needs with sun, wind and water, including making significant quantities of renewable hydrogen to power our future. Many productive careers will be built in the hydrogen space in the next few decades, and green hydrogen investment opportunities will be numerous. Some will create fortunes, others will lose them. But that's where our efforts, ideas and money should be directed.

8

Nuclear Power Is So Over

The Yucca Mountain Nuclear Waste Repository, tucked away north-west of Las Vegas in the US state of Nevada, is probably the world's biggest and most useless white elephant. Since being designated in 1987 as the intended permanent storehouse for US nuclear waste, a staggering $19 billion has been spent on its construction, including building underground pipelines buried 300 metres below the mountain to hold steel containers full of radioactive waste.

Highly radioactive nuclear waste is leftover fuel from either nuclear power plants producing electricity, nuclear weapons production sites, or plants that reprocess and recycle used power plant fuel. It remains extremely dangerous to human beings and to nature for many thousands of years, and therefore needs to be isolated, without possibility of leaking, for millennia.

The US has 94 operating and 23 defunct commercial nuclear power reactors in 34 different states (though ironically not in Nevada). They all keep their waste product above ground, in what have always been temporary facilities (even though some have been around since the 1940s), typically near the nuclear electricity generation facility it came from. Eventually all the waste was supposed to be shipped to Nevada, to be stored there permanently deep underground and solve the problem once and for all.

Yucca Mountain is millions of years old and was formed from the ash produced by ancient volcanic eruptions. Sacred to the indigenous Western Shoshone nation, its extremely high volcanic ash content and consequent ability to absorb radiation were key considerations in its selection. Over three decades later, the construction of America's one and only permanent dumping ground for hazardous nuclear waste is still not complete. Furthermore, the site has never been used and most likely won't be. The project has also been defunded by the federal government since 2010, in the face of strong opposition from Nevada's representatives to the US Congress, the Nevada legislature, the local citizenry, non-governmental organizations and the Shoshone indigenous people.

The long-term fate of US nuclear waste continues to be postponed from generation to generation and tensions flare up occasionally between the US federal government and the individual states. In a recent episode, for example, the State of Texas legislature blocked a plan to store additional radioactive nuclear waste at a site in Texas while contemporaneously the federal Nuclear Regulatory Commission was issuing a licence to a private sector company to build and operate an interim storage facility there. Texas, like other states, doesn't want to become America's nuclear waste dumping ground.

This waste challenge isn't specific to the US. No permanent repository for nuclear waste exists yet anywhere in the world. Finland is the only country with one currently under construction; if completed, this will enter service no earlier than 2025, 71 years after the world's very first nuclear power plant was commissioned in the Soviet Union in 1954.

The fundamental problem of nuclear energy isn't even the waste, even though that remains intractable. It's the fact that public opinion doesn't like it one bit, with many instantly thinking 'Armageddon' as soon as the word 'nuclear' is uttered. The industry's accidents invariably capture the public imagination because in every case they could result – though they haven't yet – in an enormously tragic loss of life across a vast geographical area. Near-apocalyptic events include the 1979 Three Mile Island accident, a partial meltdown of a nuclear reactor in Pennsylvania, the 1986 Chernobyl accident in the Soviet Union and more recently the Fukushima Daiichi nuclear disaster in Japan.

The number of operational nuclear reactors worldwide has remained stable since 1990 at approximately 450, up from 11 in 1959. In the US, one entered service in 2016 but it was the first to do so in 20 years. In the UK, no new nuclear power plant has entered service since 1995. Around the world in the 1970s, a typical future electricity forecast assumed that nuclear energy would provide 20 per cent of everything we needed globally, while the industry itself thought that figure could reach 50 per cent. Nuclear power, however, accounts for less than 10 per cent of global electricity today.

At the same time, several countries using nuclear power to generate electricity for civilian use began to build nuclear weapons, increasing the risk that nuclear weapons technology might be used as an assault weapon by a sovereign state, or end up in the hands of terrorist organizations or other armed groups. No wonder the general public is uneasy about the technology.

Every time the industry or government claimed that the nuclear power problems had been resolved – typically they would point to France, which at one point had 80 per cent of its power from nuclear energy – the industry would be hit by another disaster. There have been more than 100 nuclear incidents to date, although the information is hard to come by because even the International Atomic Energy Authority, astonishingly, keeps an incomplete historical database, with patchy accident reports covering only the period from 1988 to 2018.

At the same time, small, compact and therefore safer reactor technologies haven't worked either. After over 70 years, nuclear warships today still use reactors that are not that different from the original 1950s design, and the industry has progressed little for decades.

The nuclear industry is a glutton for subsidies. These aren't always transparent or simple to quantify, because a significant proportion are in the form of shifting construction costs, operating risks and the perennial waste-management problems. In the United States, for example, the nuclear power industry is estimated to have benefited from subsidies valued at more than the power the industry produces. In other words, buying power from any other source and giving it to households for free would have been cheaper. In the UK, Hinkley Point C, which has the dubious honour of possibly becoming the first nuclear power station in the country for 30 years if it is completed in 2026 as planned, benefits from subsidies of $55 billion even before the true costs for the long-term storage of the hazardous waste and full decommissioning have been taken into account.

All this means that households pay much more for electricity than they should, especially in the age of cheap and plentiful renewable energy.

If nuclear power was a practical solution to our decarbonization imperative, China would be the natural leader, because China has literally tried everything. Most of the country's population and almost all its industry is situated very close to enormous sources of cooling (nuclear reactors require huge amounts of water, much of which is then dumped into oceans and rivers, unregulated and causing ecosystem havoc). But plans to deploy multiple reactors along its coastal provinces have been repeatedly revised since the 1980s. China partnered first with the French, then with the Germans, and then went for it alone. They built pilots of small, safe new technologies one after another, but they still haven't succeeded in putting together a competitive offering on the global market. That's quite a rare failure for China. They're still at it to this day, experimenting with new nuclear power solutions.

For example, Beijing announced in 2021 that it had finished building a thorium-fuelled molten salt nuclear reactor in the middle of the Gobi Desert in the north of China. This would be the first specimen of that particular technology operating in the world since 1969, when the Americans shut down their pilot. This design uses molten salt and thorium as fuel instead of uranium. Many aspects of this technology attracted the Chinese. First, it is safer because the nuclear reactor can't get out of control and damage the reactor structures. Second, the reactor doesn't need to be built near water – and therefore can be away from population centres – because the molten salt itself serves as a coolant. Finally, thorium is much more widespread in nature than uranium. China in particular, rich in rare earth metals, has plenty of it.

However, there are serious problems with the technology. A fundamental one is that the thorium nuclear reactor doesn't just produce energy. It also produces uranium-233, which as it happens doesn't exist in nature and can be used to build atomic bombs. Every time the nuclear industry tries to lift its game, it creates additional problems that it doesn't know how to resolve.

The pro-nuclear lobby believe that because nuclear power is a zero-emissions energy source, we should deploy more of it faster. It's absolutely true that it doesn't produce emissions: nuclear power is generated through fission, splitting uranium atoms to produce energy to release heat to create steam to spin a turbine. No greenhouse gases are generated.

However, it is not cost-competitive. France, a country highly knowledgeable in nuclear power, thinks it needs 15 years and €20 billion to build a new nuclear power plant. Today, however, €20 billion can buy enough solar panels to generate 100 gigawatts of solar power (a little less than the installed power generation capacity of Spain), which can be installed in no time, instead of the 15 years or longer required to build and commission a nuclear power plant. Both the €20 billion and the 15 years are in any case very likely large understatements: Hinkley Point C has seen cost increases of over 20 per cent (with more on the way, according to French utility EDF, its majority investor), while a French nuclear reactor under construction has been delayed by over 10 years (and counting).

So what's the point of squandering more time and money on nuclear power? There isn't any.

Periodically, opinion articles and research pieces about nuclear power appear with headlines such as 'Nuclear power is the fastest way to slash greenhouse gas emissions and decarbonize the economy' or 'The climate needs nuclear power'.

But we know what we have to do to fight climate change. We have to stop using most fossil fuels – oil, gas and coal – by 2050. We know how to do that: we need to decarbonize our economies and lifestyles using clean and green energy, while extending electricity to the billion people around the world without any. We also have to stop deforestation and protect biodiversity.

Nuclear advocates think we should deploy more nuclear power plants faster to decarbonize the world, and in a way they're right. If nuclear power was cost-competitive, it would be a good plan. If we knew what to do with nuclear waste, some of which takes between 100,000 and a million years to lose its radioactive characteristics, it might make sense. If nuclear reactors didn't need 615 gallons of water per day to provide power for just one home – as opposed to none for solar and wind energy– or 20 years from planning to operation (versus six months to two years for solar and wind power), then nuclear power would be a great idea.

But nuclear power costs four or five times as much as solar and wind, and pollutes 30 times more. And we haven't even mentioned the weapons and meltdown risks. We don't have time to kick the can down the road and trade one set of problems, from oil, gas and coal, for another, from nuclear power. Everyone – individuals, start-ups, corporations and governments – should be focused on solar power, wind power, hydro power and tidal power in order to bring a sustainable energy future into the present as cost-effectively and as fast as possible. We're wasting time just talking about nuclear power, at least when we discuss fission.

However, there is another nuclear challenge that is both exciting and very much worth investing in. It's called nuclear fusion, and nature itself invented it. About 100 million years after Big Bang, there was a first fusion reaction, which was produced in an ultra-dense, ultra-hot core of the gigantic gaseous sphere. As a result, the very first star was born. That was followed by literally billions of other stars in a process that continues to this day.

The way nuclear fusion differs from nuclear fission, which is what powers today's nuclear plants, is that fission takes big unstable atoms and splits them, whereas fusion takes small atoms and combines them to forge larger ones. Think about a star being formed. Fusion is the universe's ubiquitous power source because it's what causes the sun and stars to shine. It's the reaction that created most of the atoms we are made of.

We have been researching nuclear fusion since the 1940s, and specifically the mind-boggling challenge of how the power of the sun and the stars can be harnessed in a man-made machine.

The beautiful thing about the fusion dream is that fusion doesn't produce carbon dioxide or radioactive waste, and the fuel it needs – certain types of hydrogen – is plentiful. A hypothetical nuclear fusion plant wouldn't take up any space either, compared to what we need for solar and wind, for example.

The dream is getting closer to being realized, too. In 2022, an experiment to reproduce the power of the stars in a box at the Joint European Torus (JET) reactor in Oxfordshire, England, smashed its own world record dating back to 1997 for the amount of energy produced during a controlled, sustained fusion reaction. That's pretty amazing, but the energy released is just enough to boil 60 kettles.

We've been on a decades-long quest to produce fusion energy and then use it to power the planet. If we succeed, it's likely to be the greatest scientific achievement ever. Nuclear fusion, in contrast to fission, is an area into which we should put very serious research and development dollars – certainly a lot more than we're doing today. It won't mean we can deliver nuclear fusion in the next 50 years (a prospect usually derided as forever 20 years away), but we mustn't stop trying. Investors are beginning to agree: a nuclear fusion start-up managed to raise $500 million in 2021 for its technologies, and another 35 private fusion companies have been launched in the early 2020s, backed by over $2 billion of funding.

Government and regulators are also on board: ITER, the $20 billion experimental fusion reactor being built by a multinational government consortium in Saint-Paul-lès-Durance in Provence, France, backed by 35 countries, is JET's far bigger successor and it is expected to become the first fusion device to create net energy, with a goal to produce 500 MW of fusion power from 50 MW of input heating power. The UK Atomic Energy Authority is pitching fusion as an eventual replacement for gas and coal – supplementing solar and wind technologies – and since 1965 has had a purpose-built institute, home to Britain's fusion research programme, in Culham, a village south of Oxford.

Conventional nuclear power is finished. Innovating in that sector is not worth time or money. Pouring more money into it is a monumental waste that diverts funds from worthier pursuits, given the urgency of our climate challenge. Renewable energy is plentiful, cheap and can be built far more quickly. It needs all those dollars that are going into conventional nuclear power research, all those dollars that are going into oil, gas and coal exploration, in order for us to deploy renewables about three times faster than present. This is the only way we'll have even a small chance of keeping temperature increases to 2° Celsius above pre-industrial times. We simply don't have the 20 or 30 or 40 or 50 years that nuclear power would need as a solution.

But at the same time, we should most certainly invest in nuclear fusion, because if we figure out how the power of the sun and the stars can be harnessed in a man-made machine, we would have solved our energy problem for ever.

9

Stinky Gas

In late 2019, the lending arm of the European Union and the largest multilateral lender in the world, the European Investment Bank (EIB), presciently said it wouldn't be lending to fossil fuels any more. The oil and coal part wasn't surprising. The gas bit shocked. That's because the world had been told time and again by oil companies that natural gas is green, clean and a bridge to a better future. The biggest multilateral bank in the world said this wasn't true, even though it had no clue a Ukraine war was around the corner.

It turns out that whoever named it 'natural gas' instead of 'highly explosive fossil fuel gas' deserved a world-topping branding award. By rebranding methane gas as natural gas, it became the fossil fuel with the most positive image, oozing fresh, organic characteristics. It was brilliant greenwashing.

Sadly, natural gas's climate impact is like CO_2 on steroids. Methane is 84 to 86 times more potent than CO_2 over a 20-year period, and when it's burned, it becomes CO_2 itself. Natural gas leaks methane throughout its entire industrial process: when extracted, processed, stored, transported and used. Methane levels are up 50 per cent in the last five years, threatening our ability to deliver on the Paris Agreement, while leakage from natural gas at 4.1 per cent means natural gas is worse than coal.

Yet the oil and gas industry managed to show that Abraham Lincoln's quote 'You can fool all the people some of the time, and some of the people all the time, but you cannot fool all the people all the time' wasn't true. They fooled all the people all the time for a very long time about 'natural' gas. But the European Investment Bank, the bank that lends the most money in the entire world, finally caught up with them. Its energy choices matter. Every one of our choices as citizens matters too, of course, at the very least because they make us feel better. Because of its size and influence, however, EIB's energy choices matter more. At a stroke, it can turn off an enormous tap of money to gas.

Historically, the EIB is far from being a climate leader – it lent more than €12 billion to fossil fuels after the 2015 Paris Agreement, and as recently as 2018, it invested €2.4 billion into one of the most expensive and controversial infrastructure projects of all time, the Southern Gas Corridor, an initiative to ship natural gas from the Caspian and the Middle East to Europe by building enormously long pipelines. Each euro invested in the Southern Gas Corridor generates €13 of climate damage. All the more reason why EIB's decision to drop natural gas was courageous: we're not investing enough in climate action and we are losing precious time getting distracted by natural gas diversions.

The Intergovernmental Panel on Climate Change (IPCC), the UN body for assessing the science related to climate change and providing governments with the scientific information to develop climate policies, says that the world needs annual investments of $2.4 trillion in the energy sector alone until 2035 to limit temperature rise to 1.5°C above pre-industrial levels. This rises to $4.35 trillion by 2030 to, in addition, decarbonize our buildings, transport systems and industries and get them off fossil fuels. But $4.35 trillion per year is a seven-fold increase on where we are now, which is approximately $630 billion. Worryingly, the rate of increase in global climate finance flows has been slowing down instead of accelerating: annual climate finance flows have been increasing at approximately 25 per cent a year since 2013, but are now plateauing, slowing to just 10 per cent in 2020.

Public finance institutions and development finance institutions like the EIB and the World Bank are doing what they can. They and other public sector actors, such as governments, provide about half of the total $630 billion of annual climate finance. The private sector is lagging, however. We can divide the private sector into four pots of money: corporations, banks, institutional investors, and households or citizens. Households and citizens are doing well. They are investing $55 billion a year into the energy transition by buying electric cars and electric boilers and putting rooftop solar panels on their homes.

Corporations, however, aren't pulling their weight. Companies like Apple, IKEA and others invested $124 billion on average in 2019 and 2020 into the energy transition, but that's a decrease from the $156 billion they invested on average in 2017 and 2018. This is consistent with findings that out of 107 companies committed to the Climate Action 100+ initiative (over 600 institutional investors globally with $55 trillion in assets engage with large companies

83

on climate), 70 per cent had no climate-related disclosure in their 2020 financial statements. In other words, they're talking the talk but they can't explain how they're walking the walk.

Banks are leading backwards. They invested an average of $122 billion into the energy transition in each of 2019 and 2020, but lent almost six times that amount, $700 billion, to oil, gas and coal in 2020 alone.

Institutional investors – asset managers – are tragically not supporting the energy transition at all: they account for a paltry $4 billion out of the total climate finance of $630 billion, or less than 1 per cent. Yet they play a crucial role in whether we manage to put the brakes on the world heating more than 2° Celsius compared to pre-industrial times. Just the largest 500 investors control more than $100 trillion in assets under management, the equivalent of 35 times the annual output of Britain's economy, or five times America's. How they channel that financial firepower can dictate how fast society can decarbonize.

The European Investment Bank's move can radically change those dynamics: it's an early indicator that the nails have started going into natural gas's coffin. The bank stated clearly that it wasn't going to invest in 20-to-25-year projects that are likely to lose their economic value over that time (in other words, become stranded assets) and that aren't consistent with the EU's climate targets.

This opens the doors to climate activists to launch more lawsuits against both producers of natural gas and their private sector lenders. If lending to the gas sector is a breach of fiduciary duty, given the potential future losses that the European Investment Bank has highlighted, then that will play out in the same way as it did in the coal sector. The EIB's move is a sign of an emerging legal obligation for others to consider the material probability that natural gas assets will be stranded in the near future.

It has been a long time coming – too long – as a case study from the Netherlands shows. Groningen is a huge gas field in the Netherlands, active since 1959 and operated by Shell and ExxonMobil. It's the tenth largest in the world and the largest in Europe. It generates natural gas equal to half of what Japan consumes. There's a long-standing problem with Groningen, however: earthquakes. When you extract natural gas, you do it from a gas reservoir. The more gas you extract, the emptier the space underground. Water then comes in, fills up the space and the ground becomes unstable.

This instability has induced almost 1,400 earthquakes since 1991.The people living nearby complained for decades but were ignored. ExxonMobil and Shell disputed the science and argued that the earthquakes had nothing to do with them. The Dutch government was making a lot of money from Groningen and looked the other way. The complaints mounted, until at last, in 2014, the government decided to start reducing output from the gas field and paying compensation to the owners of the 90,000 homes in the near vicinity that had been damaged, at a cost of more than $10 billion. Despite this, production continued and so did the earthquakes.

Meanwhile, Shell and ExxonMobil used delaying tactics to deny residents the compensation they were entitled to, stalling, shifting blame and using legal procedures to bring cases to a halt.

Finally, in 2019, the Dutch government decided that they were going to close Groningen in 2022, instead of 2030. The local community is still awaiting compensation from ExxonMobil and

Shell, while ExxonMobil and Shell are still arguing that the Dutch government should compensate them for the loss of production stemming from the early closure of the gas field.

Groningen was soon back in the news because of the Ukraine war. Proponents of fossil fuels argued that it needed to remain open in order to help wean Europe off Russian gas: about 20 per cent of the EU's total energy is from fossil gas, and 40 per cent of that comes from Russia. In the case of Germany, Europe's largest economy, reliance on Russian gas can be anywhere from 32 per cent to 60 per cent, depending on the season. If you depend on imported fuel, whether it's gas, oil or coal, the producer has leverage on you. In addition, your citizens and businesses are at the mercy of price fluctuations caused by events outside your border.

The Ukraine war made it quite clear to both the public and governments everywhere that we had a double emergency on our hands, a climate emergency coupled with an energy security emergency; and that both emergencies had the same root cause: dependence on fossil fuels.

Instead of saving Groningen, demand for gas must be cut, a task that should have been undertaken years ago. In truth, the alleged difficulties of breaking Europe's addiction to fossil gas are nonsense. It can be done, it must be done and now, because of the Ukraine war and its after-effects, we know that it is going to get done.

How? Demand for gas can broadly be split between what we use in our homes, what we use in our factories and the gas we burn to generate electricity. In Europe, the largest component is residential use, accounting for 40 per cent of total demand. All three demand sources can be tackled simultaneously.

First, our homes. We should commit to converting every building off fossils (as nowhere has done), mandate that every new home comes with solar panels (as California has done) and ban new gas pipe connections to buildings and homes (as New York City and Berkeley, California, have done). These measures are long overdue and will allow industry to develop continuously improving solutions for smaller properties and bring down costs. Meanwhile, we need to stop our dependence on gas for heating, hot water and cooking. Not only do our gas-powered appliances depend in most cases on gas piped from another country, they're also self-harming: we're literally hurting ourselves on purpose. For example, UCLA's Fielding School of Public Health found that replacing gas with electric appliances in California yields $3.5 billion in health benefits and prevents 350 premature deaths each year.

These findings are applicable everywhere in the world. The solution is readily available in the form of pollution-free induction hobs and heat pumps. Heat pumps in particular are fundamentally better products than their fossil alternatives; they operate on electricity and are three times more efficient. Speeding up their deployment would cut gas demand from housing by more than 66 per cent. But we're not manufacturing enough of them, partly because of perverse incentives to install gas furnaces – such as in Austria, which doles out a bonus of up to €993 – and partly because of the lack of qualified installers. France scrapped subsidies for new gas heater installations only in April 2022. The UK has plans to install 600,000 heat pumps by 2028, but doesn't have the required human resources to get this done. The US is even considering invoking its Defense Production Act to significantly scale up its capabilities to manufacture and install heat pumps.

More countries should follow the example of Ireland, which is enacting possibly the most ambitious home retrofitting plan anywhere in the world, with the aim of minimizing residential carbon emissions and saving people money on their energy bills. Ireland is planning to retrofit 30 per cent of its homes (that's 500,000 properties) by 2030 via a barrage of measures including installing 400,000 heat pumps; free energy upgrades for those most at risk of fuel poverty; grants of up to 50 per cent of the cost of the deep retrofit; and special enhanced grants to cover 80 per cent of insulating attics and cavity walls. Elegantly, more than half of this will be funded by money collected directly from polluters via a tax on carbon. Other European countries can tap the same source of funding, as well as the €1 trillion mobilized by the EU's Green Deal Investment Plan.

Everything applicable to homes can be and should be implemented in offices too, and climate action movements such as the UK's Insulate Britain should be supported rather than shunned: they anticipated the effects of the Ukraine war by identifying the problem (energy inefficiency caused by bad insulation and household energy poverty caused by rising gas bills) and pushing for a solution not dissimilar to what Ireland is implementing: a government-mandated and funded low-energy and low-carbon whole-house retrofit of all homes in Britain by 2030.

After homes come industries and factories. Fossil gas plays a key role in multiple industrial processes. Take fertilizer plants, for example. Fertilizer is made by converting natural gas into hydrogen and then combining that with nitrogen, an energy-intensive process with a significant carbon footprint. Eliminating black hydrogen should be a top priority, but this is co-dependent on scaling up renewable energy faster. Steps are being taken in this regard, though frustratingly only after a fossil fuel war exposed the dangers of energy dependency and forced governments' hands, while the surge in gas prices meant that green hydrogen became cheaper than the black version.

In 2022, in literally a period of just days, after years of slow-rolling renewable energy and climate action, Germany committed €200 billion to bring forward its goal of 100 per cent renewable energy by 15 years, from 2050 to 2035, tripling the pace of deployment of offshore wind, onshore wind and solar power. At the same time, the Netherlands and Italy both committed to unlocking several tens of gigawatts of offshore wind power, while the UK floated changes to its planning rules to make it easier to build wind farms. Governments are finally paying proper attention to the fact that the market price of sunlight and wind, the fuel necessary to power solar and wind power plants, does not fluctuate. It will stay at zero throughout the Ukraine war and its aftermath, and will remain at zero forever.

Much more can still be done to accelerate renewable energy deployment further and wean economies off imported gas in the shortest possible time. Europe could boost solar power deployment by as much as 50 per cent each year and wind power by 20 per cent each year compared to what is currently being planned, just by making approvals and permits more efficient and prioritizing major projects. It is ridiculous that in some European countries, developers of renewable energy projects are still required to physically print out tens of thousands of pages to bring hard copies to local planning offices.

Cutting gas use fast also means delaying nuclear shutdowns where possible, as Belgium did when it extended the lives of its two newest reactors by up to 10 years in March 2022. This is helpful in the short term because it facilitates the transition to renewable energy while not compromising on climate change.

We started 2022 with the EU reconciled to needing Russian gas for decades. This then morphed into narratives about how difficult it was to decrease dependency on Russia, before the International Energy Agency released a roadmap to cut Russian gas imports by half by the end of the year. That's quasi-instantaneous the way energy policy goes. And that was before the EU went a step further and announced its plans to cut Russian gas imports by two thirds by the end of 2022, and by 100 per cent by 2027. Crucially, only very limited supplies of gas are available from other fossil-fuel-producing countries, such as the United States, Qatar, Algeria and Australia, and their ability to supply Europe will remain constrained for several years. This means that we're not replacing Russian fossil gas with fossil gas from elsewhere. The potential to cure our addiction to fossil gas has never been clearer.

See also:
Chapter 7: Hydrogen Makes Up 70 Per Cent of the Universe;
  I Didn't Know That Either;
Chapter 8: Nuclear Power Is So Over;
Chapter 19: Going Vegan to Go Green? Don't Bother;
Chapter 28: It's Raining Renewable Energy

10

Never Buy Carbon Offsets for Anything, Especially Your Car Gasoline

Oil traders are offering their customers 'carbon neutral' cargoes of liquefied natural gas, or LNG. That is intriguing. Liquefied natural gas is fossil gas that's been turned into a liquid so that it can be transported on ships and trains and through pipelines. Once it reaches its destination, it's reversed back into the fossil gas it used to be and utilized. It can be found everywhere, in homes, offices and industries, and its applications are many, including heating, cooking, generating electricity, fuel for vehicles, manufacturing plastic, fertilizer and a large number of industrial products. It also leaks methane, the powerful climate-warming gas, at every step of the way, from extraction to processing, transport and final use. It's anything but carbon neutral. Oil traders, however, found a solution to present it as such: a cargo is sold together with carbon offsets that the traders buy to zero-out its emissions.

The carbon offsets are purchased from voluntary carbon markets and are mostly created by planting trees; then estimating how much carbon that tree is going to absorb; then printing that amount on a piece of paper (or its digital equivalent) and selling it to polluters. This is all done within frameworks that regulators and governments aren't policing and where participants are left to their own devices.

Because the voluntary carbon markets are a free-for-all, the oil traders presenting their LNG as carbon neutral say very little about how the emissions from that cargo were estimated; whether they include emissions throughout the life cycle of the gas, or just the emissions from transporting it on a ship; where the carbon offsets were purchased from; and what the oil trader paid for them. They want us to believe that the very same old dirty polluting LNG is now fine because the oil traders cleaned it.

More broadly, a vast number of companies believe they can continue with business as usual and manage some or all of their resulting emissions by offsetting these with carbon credits they acquire from people planting trees elsewhere. Everybody wins. A typical example would be airlines such as Delta, which spent $30 million in 2020 to offset 13 million metric tons of its carbon emissions that year, as part of its pledge to help combat climate change.

Oil businesses take this magic trick to the next level, announcing their intention to both increase total fossil fuel output and achieve zero emissions at the same time. Shell alone is promising to find more carbon offsets (120 million tons of CO_2) than were generated by the entire world in 2020 (104 million tons of CO_2). Hundreds of thousands of companies think they can continue doing what they do while generating carbon emissions and looking great doing it.

In reality, companies buying carbon offsets when not required by law are greenwashing. In addition, consumers and citizens buying carbon offsets on a voluntary basis – for example, to mitigate their flights or the pollution from their gas-guzzling SUVs – are being exploited and should stop buying them.

The story of carbon offsets started a long time ago. The global legal construct around climate change was born in 1992, when the UN Framework Convention on Climate Change was negotiated at the UN Conference on Environment and Development, more commonly known as the Earth Summit, in Rio de Janeiro.

Most countries in the world agreed then that we were interfering with the earth's climate by pumping enormous quantities of climate-warming gases into the atmosphere. The gases in turn interfered with our natural climate cycles, threatening more floods, droughts and extreme heat than we could handle. There was also increased risk of the Arctic and Antarctic ice melting, glaciers in the Himalayas rapidly shrinking, sea levels rising and so on.

We did not know then exactly when our impact on the climate would be everywhere for everyone to see, or how exactly this would play out. But we knew it would lead in time to cities and countries being submerged, others becoming uninhabitable and yet more experiencing food and water shortages, and many climate refugees being forced to move to more hospitable lands. We had to act.

The 1992 Earth Summit was a historic first, because it was then that the world started debating the complex relationship between economic development and environmental sustainability. In particular, we agreed that countries had what is called 'common but differentiated responsibilities' in fighting global climate change. This was code to indicate that as industrialized countries were richer and better equipped to confront the issue, they should take the lead in fighting climate change. It was a beautiful document at a beautiful moment in time in a beautiful city, and everybody signed it.

Once they left Rio, however, the wealthy nations of Europe, North America and Australia promptly ignored it. That was a foretaste of the decades of climate change talks that followed. Rich countries especially overlooked the part that clearly indicated that there was a difference in responsibility between themselves and poorer countries. Rich countries were – and still are – responsible for the overwhelming majority of the pollution sent up into the atmosphere since the Industrial Revolution two hundred years ago.

US President George H. W. Bush gave a sneak preview of the difficulties the world would face in confronting this global problem by declaring that the 'American way of life' was not negotiable. I am not exactly sure what he meant, given that clean energy and a cleaner planet actually enhance livelihoods, but I suspect he meant that Americans wanted to keep their ultra-large refrigerators, gas-guzzling cars, energy-inefficient buildings and trillions of supermarket plastic bags. Yet at the same time somehow they would decrease their dependence on burning oil and gas.

From 1995, climate talks became an annual event where countries were regularly updated on the latest in climate science, exhaustively discussed policy responses, negotiated joint efforts, adopted declarations, then promptly forgot about the whole thing. That was the story until 2005, when something called the Kyoto Protocol was ratified. The Kyoto Protocol was the first concrete attempt to move from symbols to action by presenting to the private sector the opportunity to make money while saving the world at the same time.

Pollution is agnostic. It does not matter where it is generated: it all ends up in the same place: our planet's atmosphere. The Kyoto Protocol was a courageous effort to attack the problem globally. Its aim was to reduce pollution everywhere through market-based mechanisms and pollution reduction certificates, also known as carbon credits, carbon allowances or carbon offsets. This market-based method had proved to be an unquestionable success in the United States in the 1990s, when the 1990 Clean Air Act included pioneering cap-and-trade plans. These created markets that attacked sulphur emissions (generated by fossil fuel combustion at power plants and resulting in harmful chemicals literally raining on people and places), and within a few years, the entire problem was gone.

This precedent established that market-based mechanisms worked, provided they were regulated and enforced. The Kyoto Protocol, however, had two fatal flaws. First, the biggest polluter in the history of the world, the US, never ratified it: President Clinton signed it, but in his first year in office, George W. Bush pulled the United States out. Second, the next biggest emitter of greenhouse gases, China, wasn't included in the Kyoto Protocol as a polluter because of its status as a developing country.

Today there are many markets for carbon offsets, which can be broadly divided into two: those mandated by laws and regulations, and those that are voluntary, where nobody is actually asking anybody to do anything. The carbon markets mandated by law are called compliance markets and are governed by rules. The biggest and most famous are the European Union Emissions Trading Scheme, California's Greenhouse Gas Scheme and the Chinese Emissions Trading Scheme. New Zealand has one as well, as do some north-eastern US states. There are other pockets of regulated carbon markets elsewhere.

In the compliance markets, if you're a business in certain sectors and you emit greenhouse gases, you're required to either decrease your emissions if they exceed a certain level (a cap), or alternatively buy a credit for each ton of carbon dioxide equivalent that you emit annually above that cap. You can buy these credits from the government or from businesses who reduce their emissions. If, on the other hand, you have managed to cut your emissions through, for example, better energy efficiency or capturing methane or many other techniques, you can earn carbon credits and then sell them to somebody who is polluting beyond their cap. Periodically the cap is decreased by regulators and in time the pollution goes down to zero.

This mechanism commoditizes carbon credits and creates a market between all the companies to which the regulations are intended to apply (typically one or more sectors of the economy, but not all). These markets generally trade only in their own carbon allowances, although some accept those that come from a pollution reduction activity in a different country that complies with strict rules. These compliance markets are organized and regulated. They're proven to work. The only key to how well they work is political will to maintain a steady and increasing carbon price over time. There is no other challenge.

These regulated carbon markets, including the Kyoto Protocol and the European scheme, took off after 2005, followed by California's in 2013 and China's in 2021. The Kyoto Protocol market, however, subsequently collapsed and disappeared because of a lack of unified global political will. Other markets, where political will did exist, are working increasingly well at regional, country or state level, and are delivering world-leading results, in particular in the EU and California.

At the same time as these regulated carbon markets took off, a voluntary market turned up as well, one in which no government or regulator was telling anyone what to do. Voluntary carbon markets set their own standards, which are used to certify that a carbon credit is respectable. Developers go out and create projects such as afforestation (planting new forests), turning around degraded forests and reforestation. Companies started buying these carbon offsets in order to voluntarily meet their carbon reduction goals.

Back in 2005, the expectations of market participants were that this voluntary market would run until 2020 only. We knew then that come 2020, we wouldn't have any more time for solutions that allowed companies to continue to emit, as the voluntary carbon market allows. We understood that the concentration of carbon in the atmosphere was increasing each year, and that the rate of increase meant that by 2030 we would have locked in irreversible climate change at very dangerous warming levels. Today, we have to move not to net zero, but to zero. Therefore actions that do not decrease emissions directly are meaningless.

Ironically, not much had happened in the voluntary carbon markets for many years. But in 2020, they came roaring back. The supply of carbon credits was up 76 per cent as forestry projects became hugely popular and big companies started buying their carbon offsets. That's because smart people at polluting companies around the world figured out how to do very little yet still look amazing. 'Vitol is selling carbon-neutral LNG' sounds good. In reality, it's meaningless. The carbon offsets attached to the LNG are coming from unclear sources; aren't supervised by governments; aren't subject to any laws; are certified by unregulated bodies and monitored and verified in a haphazard manner. In addition, they shouldn't have been issued to polluters in the first place because it takes years to know that the trees they represent have stayed around long enough to make a difference.

Here's a case in point. The 2021 California wildfires fires burned through at least two forestry projects that had sold carbon offsets to several companies, including BP. As climate change increases the frequency and severity of these fires, large quantities of carbon offsets will literally go up in smoke.

The entire voluntary carbon market is an illusion. LNG traders and other businesses seeking to buy carbon credits to offset their emissions are welcome to go and buy compliance offsets, i.e. offsets that are regulated by the European Union or by California. But they won't, or don't, because it would cost them dearly. Companies are paying \$2 or less for voluntary carbon credits that represent one ton of CO_2, whereas prices on regulated markets in Europe are twenty times that amount or higher.

In fact, if oil and gas companies had to buy European compliance offsets, many would be technically bankrupt in short order, because regulated offsets are starting to communicate the real price of burning oil, gas and coal and sending a signal that says we can't afford it any more. It's time for companies to spend their money on doing the right thing, which is decreasing emissions from their operations and their supply chains – especially if they commit to do so via announcements about their carbon neutrality goals. They are greenwashing unless they are either decreasing emissions or buying compliance credits.

Many individuals want to do the right thing and decrease their own emissions. They might therefore decide to calculate their carbon footprint from eating and drinking, travelling or investing, then buy carbon offsets to make themselves carbon neutral. They should absolutely not do that.

Shell, in the Netherlands and the UK, gives car and truck drivers the option to add one cent (or one pence) per litre to the price of their fuel at the pump, which the company says will be used to offset carbon emissions. According to Shell, the money will go towards buying carbon credits from nature-based projects: code for planting trees. It assures consumers that it is investing $300 million over three years to make this happen, and its public relations machine is in overdrive to market this initiative. Shell presents this as a great endeavour because it is giving motorists the options to drive carbon neutral for a negligible fee.

When driving, however, a litre of petrol emits two kilos of CO_2; in other words, 500 litres emits one ton. At the Shell price of one cent per litre, drivers are paying five euros per ton of CO_2. Shell, however, is paying no more than two euros for these credits, then on-selling them to their customers for two and a half times their cost. The result? Shell just turned their $300 million investment in nature-based solutions into $750 million and simultaneously pushed responsibility for their fossil fuels onto Dutch and British motorists.

That's the world upside down. Shell should instead have invested in respectable projects that generate regulated carbon credits, then it should have retired (or cancelled) these carbon credits in order to take responsibility for a small fraction of the damage the company's products are doing to the world.

But carbon calculators of this kind were invented for precisely that purpose – so companies can avoid responsibility. The very first popular carbon calculator was produced by BP because the company and its peers wanted to shift responsibility onto consumers. This allows them to increase production because it's now our fault if we consume more of their stuff – as if, for so many years, we've had any other choice.

We should not be encouraging them. Do not offset your carbon footprint. As individuals, we shouldn't have a carbon footprint to begin with. The only reason we do is because of ineffective government action, obfuscation by Big Oil and collusion by corporations over at least four decades.

Consumers are being offered voluntary carbon offsets that are both overpriced and ineffective (75% of carbon credits may be ineffective, according to research by the Stockholm Environment Institute). Here's a Bloomberg News headline that says it all: 'How Mexico's Vast Tree-Planting Program Ended Up Encouraging Deforestation: Sowing Life, AMLO's [Mexico's] flagship environmental project may have resulted in the loss of forest cover nearly the size of New York City'.

Let's take a step back and unpick what happened. Every time someone runs an engine fuelled by oil, gas or coal, CO_2 is emitted. That greenhouse gas has to go somewhere. Either it stays in the atmosphere or it goes into the oceans, to be released later into the atmosphere. CO_2 and other global warming gases that we produce alongside it (like methane, for example) prevent the earth's natural cooling cycle from working because they trap the heat near the surface of the planet, which causes global temperatures to rise, with devastating effects.

Greenhouse gas emissions have already led to a rise in global temperatures of at least 1° Celsius. We're 100 per cent locked in for a 1.5° Celsius rise, and worse is probably also pretty much guaranteed. We know this because when we talk about carbon dioxide, we talk about it in parts per million (PPM). This tells us how many parts of carbon dioxide there are in one million parts of air, so if we say that carbon dioxide is at 400 PPM, for example, this means that in one million particles of air, there are 400 particles of carbon dioxide.

It took us 5,000 years to increase the global average atmospheric carbon dioxide from 270 to 280 PPM. In other words, for over 5,000 years, nothing happened. Then, in about 1,000 years, we moved from 280 to 290 PPM as the accelerating trend began to set in. Carbon dioxide increased from 290 PPM in 1877 to 300 PPM by 1911, a period of just 34 years. The rate kept accelerating as we kept pumping CO_2 into the atmosphere. It took just five years to move from 400 to 410 PPM.

The planet is healthy for humanity at a level of 350 PPM. Right now, we're at about 420, up 10 PPM in three years, even though we've just been through a global pandemic that slowed the rise in emissions. This is the first time that our planet's atmosphere has been at that level since before modern humans existed. The last time the atmosphere contained this much CO_2 was more than three million years ago, at a time when global sea levels were several metres higher than they are today and Antarctica was partly blanketed in forests.

We are now moving fast towards 450 PPM, the level of atmospheric CO_2 concentration that corresponds to global heating of 2° Celsius above the pre-industrial levels. We add approximately 3 PPM each year, so to reach 450 will take just 10 more years.

In a world that is 2° Celsius warmer, we're going to have 25 per cent more hot days, 25 per cent more heat waves – with 40 per cent of the global population exposed to at least one severe heat wave every five years – more wildfires, more droughts, worse floods and more extreme weather events like cyclones and typhoons. If temperatures rise by three or four degrees, we would then enter a 'hothouse' stage that would make many parts of the planet uninhabitable.

Even though we've been talking about climate change for 40 years, our global collective action to date has resulted in less than nothing: we haven't decreased emissions and we haven't even slowed their rise.

The world agreed in Rio in 1992 that climate change was a huge problem.

Emissions kept rising.

Global climate talks then delivered the Kyoto Protocol in 2005.

Emissions kept rising.

The world tanked the Kyoto Protocol in 2012.

Emissions kept rising.

The world woke up again in Paris in 2015 and produced the Paris Agreement.

Emissions kept rising.

Except after the Paris Agreement came into force, companies started feeling the heat. To continue business as usual for as long as possible, they adopted a meaningless tagline, 'Net zero by 2050', which allowed them to hide behind voluntary carbon offsets.

Emissions kept rising.

A global pandemic hit. Energy use decreased dramatically, as did industrial activity and flights for a sustained period of time.

Emissions kept rising.

The world courageously congregated in Glasgow in 2021 in the midst of a global pandemic.

Emissions kept rising.

The existence of the voluntary carbon market in 2022 is a testament to our failure. This needs to be cancelled so that companies have nowhere to hide and are forced to run faster. Much, much faster.

See also:
Chapter 11: Please Don't Plant Trees;
Chapter 23: The ESG Con

11

Please Don't Plant Trees

The Amazon is many things. It's 40 per cent of South America, stretching across eight countries. It's a mosaic of ecosystems and vegetation types. It's a forest rich in biodiversity where it consistently rains. In fact, it's the world's biggest rainforest, larger than those of the Congo Basin, Indonesia, Malaysia and Australia combined. It's about the size of the United States. It's home to the most voluminous river on earth, which, during the high-water season, discharges the equivalent of 7.2 million Olympic swimming pools every day into the Atlantic Ocean. It's named after the Amazons of Greek mythology, a race of fiercely independent female warriors. It's also been burning since the 1890s.

At first, we burned it slowly. The initial large-scale deforestations were undertaken to establish rubber plantations after Charles Goodyear invented durable pneumatic rubber tyres and these began being adopted all over the world. Then we burned it fast, starting in the 1970s, when Brazil embarked on its agricultural colonization. In just two decades, we lost

18 per cent of the Amazon, frantically clearing it for farming at a rate of one football field per minute every day. We're still burning it today, month after month, losing another 1 per cent approximately every three years, the rhythm of deforestation accelerating and decelerating depending on the political mood in Brazil.

Forests still cover approximately 30 per cent of our land area globally, but they are disappearing at an alarming rate. In the last 20 years, for example, we destroyed an area of forest the size of South Africa. Since humans have been around, we've chopped down 50 per cent of all the trees.

In part because of rampant deforestation, companies have been relentlessly advertising billions of dollars' worth of plans to invest in trees. As discussed in Chapter 10, in 2019, Shell pledged to invest $300 million in what they call 'nature-based solutions', code for using the natural world to allegedly remove harmful emissions from the atmosphere. BP too is threatening to invest a lot of money in planting trees.

In 2020, Saudi Aramco, the largest oil company in the world, promised to plant two million mangroves in Saudi Arabia, a desert country without a single river. They then turbo-charged this in 2021 into plans to plant an incredibly improbable 10 billion trees. Italian oil company Eni claims that it's going to plant 20 million acres of forest in Africa. That's more than the size of Ireland. It's a lot of trees. TotalEnergies, not to be outdone, is throwing $100 million a year into reforestation. Many other companies are in on it too. I smell a rat: are oil and gas companies running around Africa, Latin America, Asia and Western Europe planting trees? Really?

Trees are about a quarter carbon. This means that as they grow, they trap (or bind) something like four times their weight in carbon dioxide. In the struggle to capture the emissions humans generate from burning oil, gas and coal and avoid these going up to the atmosphere or being temporarily stored in our oceans, forests of trees are the only technology we have. They absorb the CO_2 we exhale and trap the greenhouse gases we emit. On the flip side, when we burn trees, we release all the carbon dioxide they contain, adding to our global warming problem.

Trees are often called 'the lungs of the earth' because they absorb pollutants and improve air quality, acting like a filter for our pollution. They also produce oxygen, which we need to survive. In addition, for most people their mere existence also reduces stress: simply sitting and looking at trees helps reduce blood pressure and adrenaline, cortisol and norepinephrine, the three major stress hormones. Trees also enhance property value and provide flood protection.

If you plant a tree that otherwise would not have been there, nature absorbs more CO_2 emissions from the atmosphere. You can compute the size of these emissions, then create a carbon offset for each ton of CO_2 that is absorbed. Oil companies and others are on the bandwagon for precisely that reason: trees are their absolution. As they continue to burn coal, oil and gas, they aim to offset the CO_2 from fossil fuels with the CO_2 absorbed by the trees they've planted.

In some ways this resembles medieval indulgences. From the twelfth century until the practice was outlawed by the Catholic Church in 1567, the process of salvation was often linked to money: a sinner could be pardoned by buying an indulgence blessed by the Church, such as contributing to the building of a cathedral. Climate change, however, is not the medieval Catholic Church. You should not get to buy an indulgence for planting trees, then go back to pumping emissions into the atmosphere, exacerbating the climate change catastrophe.

I can't deny that we need more trees. We do. Trees can provide 20 to 25 per cent of the climate mitigation necessary to meet the goals set in the Paris Agreement, depending on which study you read. Many countries are trying to reverse the deforestation scourge.

Costa Rica, for example, managed to increase its forests from 21 per cent of its land mass in 1986 to over 50 per cent today, while at the same time its growth accelerated and its GDP per capita more than tripled. Ethiopia is trying to do something similar. They want to reverse a decline in their forest coverage, from pretty much zero today, astounding as that may sound, to 35 per cent, which is what it was in the early twentieth century. To motivate and mobilize their population, they began by planting 350 million trees in one day.

In Africa, a project named the Great Green Wall will, when completed, become the world's largest living structure. It's a 7,000-kilometre barrier stretching from Senegal in the west to Djibouti in the east, intended to stop desertification through a mosaic of green and productive landscapes that will fight land degradation. The project received a shot in the arm in 2021 through $19 billion of new pledges from donors including the African Development Bank, the World Bank and the European Commission. These funds are expected to be spent on restoring 100 million hectares of degraded land and creating 10 million green jobs by 2030.

Then there is Pakistan. Under Imran Khan's leadership, the country had already planted an astonishing 1 billion trees by 2017 before they decided to scale that up to 10 billion by 2023. They're clearly going to need funding for that, and they're proposing an innovative approach whereby Pakistan's creditors would forgo some of the money owing to them in return for the country delivering on its tree-planting campaign.

There's another fantastic example coming from China, which is planting forests on a dizzying scale. The country has planted an area as big as the Philippines or Finland in five years, and is single-handedly responsible for 25 per cent of all the world's new greenery since 2000. India, meanwhile, is pledging to increase its forests by a massive one million square kilometres, or two and a half times Finland, while in the space of 10 years, Western Europe has planted an area larger than Switzerland.

I did not mention Brazil, because there the destruction of the Amazon is continuing at pace. I also did not mention the United States, which isn't really doing much yet in a concerted way. Russia is the third member of that club, losing forest cover at a rate of 5.6 million hectares each year. These three countries account for 40 per cent of all global forests, and what they do at a national scale matters enormously in the fight against climate change.

But let's get back to corporations causing environmental harm and trying to mitigate that by planting trees. There are several fundamental problems with this. The first is that it provides no incentive to stop the pollution in the first place, but rather provides an excuse to continue it. The carbon offset is a penance and the environmental harm is not avoided. The second is that even if it were possible to be absolved of environmental destruction by planting trees, surely corporate polluters should not be allowed to sell that absolution? Yet that's exactly what many are doing. Shell is by far the biggest culprit: as we saw in the previous chapter, it asks consumers to pay a premium at the pump to buy carbon offsets and makes a profit from them.

What companies really need to do is straightforward: stop emitting carbon. They need to cut their emissions to zero, or as close to zero as possible, between now and 2050, and have a clear, transparent and verifiable plan of how they are going to get there. Those that can't – for example, companies entirely focused on generating, transmitting or burning coal, gas and oil – should be made to focus through shareholder pressure on downsizing their business by stopping new projects and managing a gradual, organized decline in their existing production.

Downsizing fossil businesses, counterintuitively perhaps, creates more jobs if oil, gas and coal companies transition to renewable energy (and even if they don't). To begin with, energy sector jobs tend to require the same skill sets. Offshore drilling specialists, for example, can be retrained in a straightforward manner to transfer to offshore wind farms. Electricians and power plant operators are needed in both industries. Traders can trade clean energy just as they can trade electricity generated from gas or coal. The job creation benefits of transitioning to clean energy are substantial: each $1 million we shift from fossil fuel power to renewable energy creates a net increase of five jobs.

Employment in oil, gas and coal has in any case been shrinking for years, driven by mechanization rather than climate imperatives. The coal industry employs fewer people today than in 1980, even though we are producing more of it. The contrast with renewable energy is stark: renewable energy employed 12 million people in 2020, directly and indirectly, and this is expected to grow to 38 million jobs by 2030 and 43 million by 2050.

There's a third fundamental problem with the corporate polluters pledging to plant trees. Nowhere near enough attention is being paid to which trees are being planted, how they're being planted and why. Instead, the money from Big Oil is muddying the good that others – genuinely trying to fight deforestation – are doing. The science is clear. If you want a healthier global forest, prepare the ground, acknowledge your limitations and then get out of the way. Forests are a living thing and they are far more qualified than us to reclaim the land.

By way of background, planting trees is a modern development. We did not begin doing it on a large scale until the last century. We started burning wood for fire 750,000 years ago, then used it for construction, as agricultural tools and as weapons of war. True, the Romans promoted forest conservation and reforestation, importing tree seedlings to France, Spain and Italy to replicate the pine forests of Lebanon and the groves of Carthage. But the destruction of forests continued uninterrupted until the sixteenth century, when forestry management was born in Germany to separate timber harvesting from regeneration. Elsewhere, however, forest destruction continued, to the extent that today, virtually all of Britain's woodland consists of new plantations rather than ancient forests. Forestry schools had to wait until 1825, and only in the twentieth century, when demand for wood skyrocketed, did we start planting trees in an organized fashion, because just cutting them down was no longer enough to meet our ever-growing needs for wood and paper.

Today, as long as the timber industry is properly regulated, replanting trees and managing the ratio between the trees it harvests and those it replants, it is sustainable. However, planting trees that grow quickly so that they can be cut in a few years is not an appropriate way to fight back against deforestation: we need the trees there permanently. The key is to mimic natural regeneration, which impacts how to plant, the distance between the trees and the optimum mixture of species. Even these natural replanting efforts still need to be taken care of: sites have to be prepared properly in order to ensure the trees have a good chance of growing.

In essence, what the timber industry has taught us and what the forestry scientists have shown is that to mimic natural regeneration, there's actually no need to plant trees most of the time because they do an amazing job of growing back on their own. When ecologists in the US charged with reforesting cleared grass and planted trees, versus just clearing the grass, they found that tree planting was not necessary. Five years after harrowing the grass, the test plots were full of native trees: left to their own devices or prodded by thoughtful national reforestation programmes, forests expand.

It's much more productive to first evaluate the potential for natural regeneration and then eliminate barriers to that natural regeneration. Tree planting is not a silver bullet for climate recovery, because planting for planting's sake doesn't work and a lot of the trees won't survive. Meanwhile, somebody might have earned carbon credits from these and got away with pretending to have decreased their emissions.

By far the best way to offset carbon emissions from fossil fuels is not to dig them up and burn them in the first place. Prevention is better than cure. The next best way is looking after our forests – the one proven and scalable carbon capture technology we have. To do that, the number one priority is to stop deforestation and protect ancient woodland, and be wary, very wary, of the 'I am planting trees' propaganda from polluting companies.

That doesn't mean people shouldn't plant a tree in their garden if they have one. Of course they should. It's noble and it's healthy, but that's not what we're talking about here. Meanwhile, they should never pay the extra cost of a carbon offset that has questionable impact and is almost always offered by companies that can't be trusted.

12

Have as Many Babies as You Like

D id you catch the recurring chorus in town? It's claiming that 'a newborn child will make climate change worse / it's just another soul that worsens our emissions curse'.

Mostly a Western phenomenon, a loud contingent of young people are either promising not to have kids or considering not having more. For now, they are mostly in Britain or the US. Their movements revolve around Facebook and have names like BirthStrike and Conceivable Future, and some are taking action. Advocacy movements Fair Start and Population Balance filed complaints with the United Nations on behalf of two plaintiffs, alleging that the US hasn't protected the rights of younger generations to safely and sustainably have children; and the UN has failed to protect human rights.

The issue has high-profile backing from American politician Alexandria Ocasio-Cortez, who took to Instagram to say that young people can legitimately wonder whether to have children who might grow up on a planet already at its hottest for at least

12,000 years and guaranteed to get even hotter.

According to this logic, having no children not only prevents these non-progeny from experiencing worsening conditions on earth, but also helps our fight against global warming by saving the carbon emissions the kids would generate during their lifetime.

Hold on, are we comparing having a child to buying a gas-guzzling SUV? It would seem so.

This scaremongering echoes the theories of Thomas Malthus, who, while not always wrong, widely missed the mark with his eighteenth-century theory that the world was heading for trouble because population increased geometrically compared to food supplies, which he thought increased only in a linear fashion. It turned out that we could exponentially increase food supplies too – problem solved. Although 800 million people still go hungry around the world today, it's not because of lack of food production, but primarily because of poor access to water and healthcare, as well as conflicts.

The no-babies crowd should understand that they have the world upside down, with individuals taking responsibility for actions they have no influence over instead of placing the blame where it truly belongs: with the companies that showered us with fossil fuels when they knew the harm these caused, and the governments that aided and abetted them. Who is to say that the world is over-populated when we know we can feed, clothe and provide energy to everyone without stressing the planet if we do so sustainably?

It would be far more productive to take action to change things top-down and systemically, rather than through misguided action that might cause unquantifiable self-harm. We don't know the future psychological impact on those who decide not to have children on climate action grounds.

Children are not SUVs. We can enact laws to ban these gas-guzzlers, replace them with electric cars or with no car. While we can subjugate our innate biological need to procreate and therefore proactively decide not to have children, most people will want a child because they've been raised to want one; or because they like children; or because they can think of no more fulfilling feeling than to love a child; or because they need support in their old age; or simply because of the joy, meaning and hope children provide.

This debate should never have been initiated in the first place, because it appears that sometime over the next few decades, the global population will begin to decline even without people's well-meaning interventions. We experienced sustained but very low population growth for the 10,000 years before 1800, when the rate of growth accelerated tremendously and we multiplied from approximately one billion people then to almost eight billion today. Now, however, it appears that the earth's population is stagnating, with ominous early signs from the US and China, which are both showing the slowest rates of population growth in hundreds of years.

Proponents of not having children in order to manage climate change risks are also falling into a trap set by oil and gas companies over decades to convince us all that we are responsible, through our individual choices, for the climate catastrophe. BP pretty much invented the notion of a personal carbon footprint, and popularized a carbon footprint calculator for individuals, backing its efforts with millions of dollars' worth of advertising. The campaign was labelled 'Beyond Petroleum' – subtly implying that in order for BP to move 'beyond petroleum', only individuals cutting their carbon footprints mattered.

This 'shift the blame' strategy has been taken to a manipulative extreme by Indian coal giant Adani, which continues to try to increase its coal generation capacity and is executing ambitious plans to build six new coal-fired power plants. It is aware that many countries are phasing out coal because of its inefficiency and high emissions, among them Canada, the UK, France, Italy, the Netherlands, Portugal, Austria, Ireland, Denmark, Sweden, Finland, Hungary, Slovakia and Greece. To greenwash its way back into people's good books, it launched a cynical campaign focused on children and the actions they can take to fight climate change, suggesting for example that they should cut their consumption of meat, buy recycled clothes to reduce waste in landfills, wash clothes in cold water, switch to energy-efficient appliances and use recyclable grocery bags, all while Adani itself is frying the planet with its new coal plants. I have an idea for the company: leave the children alone and instead cancel your coal expansion plans.

Harvard University research in 2017 and 2021 also showed how ExxonMobil has faced accusations of trying to shift the blame for climate change to consumers using tactics reminiscent of those of the tobacco and firearms industries. Harvard science historians Naomi Oreskes and Geoffrey Supran argued that this blame-shifting was by design. They analysed 180 ExxonMobil documents over almost 50 years – from 1972 to 2019 – to show that the words most associated with either 'global warming' or 'climate change' were 'risk' and 'demand'. The overuse of the word 'risk' was designed to make people feel that climate change was something way out in the future rather than a real and present danger, thus subtly introducing uncertainty about climate science into public discourse, while the recurrence of the word 'demand' was intended to shift the responsibility for climate change to consumers and away from the suppliers of oil and gas. Out was 'fossil fuel combustion'; in was 'energy demand of consumers'. In was helping customers 'scale back' their emissions of carbon dioxide, and asking them to 'show a little voluntary can-do', while oil and gas growth was presented as inevitable with taglines such as 'fossil fuels must be relied upon to meet society's immediate and near-term needs'. This manipulation of public opinion through the use of highly selective language was backed by hundreds of millions of advertising dollars.

In too was presenting ExxonMobil as a saviour, stating as a fact that the 'increasing prosperity in the developing world [will be] the main driver of greater energy demand' – implying that its products would also help poorer countries on their way to success. All the while, however, the company acknowledged its own culpability in climate change in internal memos dating back to 1982.

Every individual who is not having kids specifically because they might contribute to emissions in the future is falling squarely into BP and ExxonMobil's propaganda trap. It's the same trap that seeks to shame us into flying less and recycling plastic in order for Big Oil to avoid talking about decreasing their own emissions and ultimately cutting their production of fossil fuels. At the same time, it is designed to find new ways of hooking us to more fossil fuel consumption while delaying action on substitute products. Political leaders also fall into the propaganda trap, buying the plastic recycling myth, for example. The strategy is distraction – and climate babies is just one of its latest symptoms.

Individual lifestyle choices are important, and informed citizens will make them. However, time is of the essence in addressing climate change, and national and large-scale emission reduction strategies out-muscle what we can do at an individual level by a big margin.

Top-down systemic climate action is possible, and it works – we just need a lot more of it. German courts have illustrated this point vividly. In 2019, Germany enacted a law committing the country to cut its greenhouse gas emissions by 55 per cent by 2030, compared to what they were in 1990. A group of young people then filed a lawsuit arguing that the 55 per cent target was insufficient and violated human rights by shifting the burden of cutting emissions far into the future, when subsequent generations would have to live with the consequences. Germany's top court agreed, and the government was forced to move to implement a far more ambitious emissions reduction strategy, cutting emissions by 65 per cent by 2030, 85–90 per cent by 2040 and 100 per cent by 2045. These new targets immediately set in motion plans to massively decarbonize industrial plants and processes in order to cut emissions not just from the electricity sector, but also from the country's mighty car manufacturing industry, from agriculture and from its steel and cement manufacturing base, while greatly expanding Germany's renewable energy deployment plans.

No amount of individual action can deliver this magnitude of emission reductions in such a short time frame. Moreover, Germany's action shows we can eliminate all emissions if we have the political mettle to do so. Babies played a critical role in that debate, but only because we were going to have them and they deserved a better future.

13

Ride a Bicycle, Save the World

The famed Champs-Élysées in Paris was developed in several phases, starting in 1667. Created for King Louis XIV as an extension of one of his palace gardens, it didn't take its current shape until 1836. Eventually the car dealers appeared, and more than 22 automobile manufacturers were already showcasing their wares there in 1909, as the Champs-Élysées emerged as a major artery of the city. Before the onset of the coronavirus pandemic, its eight lanes were busy with, on average, 3,000 vehicles an hour. Its air pollution skyrocketed and reached levels above Paris's ring road. Complaints from Parisians grew louder, but to no avail.

Just a couple of weeks after Paris emerged from a strict three-month COVID-19 lockdown, its mayor unveiled a dramatic blueprint of urban innovation for the Champs-Élysées, taking it back to its roots as an 'extraordinary garden'. Half the space currently used for vehicles is to be replaced by cycle lanes, pedestrian and green areas and tunnels of trees to improve air quality.

The Champs-Élysées facelift is part of an ecological manifesto for the French capital, anchored around 160 km of 'coronapiste' bicycle lanes to make the entire city bikeable before it hosts the 2024 Olympics; an 'RER vélo' replicating the city's subway above ground in over 670 km of safe bicycle lanes; and the permanent removal of over 70 per cent of on-street parking spaces.

This revolution in urban planning was executed because of the coronavirus pandemic and the light it shone on air pollution and its death toll. But what has really allowed it to happen is a powerful force in our arsenal of anti-virus weapons: the humble, cheap and clean bicycle, and its twin, the electric bicycle, the perfect complements to any aspiring garden city that wants to clean up its air and enhance the well-being of its citizens.

After the coronavirus outbreak, I saw bicycles all around me. My millennial son bought one to increase his mobility. I fixed one to get it back on the road. It seemed that city dwellers had raided the world's bicycle shops. Bikes were periodically out of stock worldwide for most of 2020, as the global supply chain tried to adapt to a sudden and sustained rise in demand. New cycling enthusiasts figured out that the humble bicycle allowed them to save money, effortlessly achieve social distancing, live healthier lives and keep moving – all at the same time.

City governments around the world got on board fast, rediscovering the bicycle to get people moving without pollution and with safe distancing. Europe spent over a billion dollars on cycling infrastructure within a year of the start of the coronavirus pandemic. London, Berlin, Amsterdam, Brussels, Budapest, Madrid and Milan all created miles and miles of new bicycle lanes on streets where cars had previously been dominant, together with 'slow streets' where cars must respect extra-low speed limits and mingle with cyclists and pedestrians.

In the Americas, Mexico City, Bogotá, Philadelphia, Minneapolis, Denver, Vancouver, Winnipeg and Calgary all went big on bicycles and bike lanes. Houston and Los Angeles, two sprawling metropolises where very few people cycled to work before coronavirus, saw what promise to be fundamental changes to their urban architecture. In both cities, cycling trips more than doubled and continue to increase. New York City and Chicago were not far behind.

The urban revival made possible by the bicycle is now a broader movement that includes more green space, more car-free zones, more disability-accessible public transport and a huge expansion of subways, electric buses and trams. Given that 60 per cent of us live in cities, that's a big change. Potentially very big. Missing in action were the crowded and extremely polluted cities of Asia. Riding a bicycle in Mumbai or Ho Chi Minh continues to be an extreme sport, and cleaning the air in these places is more likely to happen via electrifying their dominant form of transport, the scooter. These are coming. In India, a carbon-negative electric scooter factory, powered by its own solar rooftop, will deliver 10 million electric scooters from 2022.

Electric bicycles are also coming to the rescue. These are two-wheel bikes that boost the rider's pedal power with a built-in storage battery. Invented in 1895, they remained a luxury item for almost 100 years, of little interest to most because of their high cost. Then they took off in China, when in the late 1990s pollution led Shanghai's government to crack down on gasoline scooters and promote the use of e-bikes. One of their greatest advantages is the fact that no new infrastructure is required for them: charging an e-bike means simply plugging its battery into a standard electrical outlet. With increased mass production, the price fell, and today there are more than 150 million e-bikes on the roads in China.

The movement has gone global. In Amsterdam in 2020, more e-bikes were sold than traditional bicycles for the first time. In the history of urban cycling, what happens in Amsterdam doesn't stay in Amsterdam. E-bikes are already on track to outsell cars in Europe later this decade, while in the United States, the growth in sales is outpacing that of regular bikes.

Electric bicycles are amazing. Like normal bikes, they come loaded with a free gym membership, because they have to be pedalled to get their motors to help. They solve noise and air pollution problems in cities at a stroke, because 35 to 50 per cent of e-bike trips are substitutes for car trips. Imagine the impact on public health if we used petrol and diesel cars 50 per cent less because e-bikes became widespread. The advantages of electric bicycles don't stop there. They cost less to buy and maintain than cars. They require fewer resources to manufacture and the city infrastructure they need costs much less than that for cars. We could, for example, get rid of half our parking in cities, useless space that we could use so much more intelligently.

The Great Plague of 1666 provoked a scarcity in labour, prompting a scientific reawakening that eventually led to the Industrial Revolution and a transformation of energy sources, driven by the burning of coal. The birth of the steam engine and of electricity followed, as did cars, trucks, buses, trains and planes. Before long, air pollution skyrocketed, as did emissions of greenhouse gases into the atmosphere, which then began to change earth's climate.

After coronavirus, in no small part with the help of the bicycle, we have started the process of reversing pollution in our cities. While the ongoing modern industrial revolution is motored by the biotech industry, big tech, big data, robotics and artificial intelligence, weaving a web of exponential technological change, the modest bicycle will play a large role in reimagining cities.

This urban planning revolution is upending multiple economic sectors and has the potential to deliver change that cleaner air advocates could only dream of. For example, parking companies will shrink in size; there will be fewer cars on city streets; city property will lose value relative to the countryside; commercial property prices will stop rising every year; oil demand will be decimated, and productivity gains due to reclaimed driving hours alone will add trillions to gross domestic product.

'Peak car year' was likely 2019 – before the onset of the COVID-19 pandemic – because of the convergence of multiple trends. These include the rise of ride-hailing; gridlock in megacities, where a car can be an expensive inconvenience; concerns about pollution; young people turning away from cars in droves; major cities restricting access to vehicles; and the expansion of the sharing economy.

Riding this wave, the bicycle-centric urban revolution will cut the space available for cars in cities, as well as the number of cars. The average car often only carries a single person, and most of the day sits totally empty, unused, simply taking up space: there are some four parking spaces for every car in existence. Imagine the waste. If you add it all up, many cities devote 40 to 60 per cent of their area to vehicles, most of which are idle more than 90 per cent of the time. As more cities emulate Paris and withdraw space from cars, cars will have to be used more efficiently and we'll buy fewer of them, parking less and cycling or using public transport more.

Shared, electric and autonomous vehicles will become cheaper, and when you're not on your bicycle, you'll buy a ride instead of buying a car, ordering on demand via an app. Autonomous vehicles are being tested on public roads around the world, and countries including Australia, the UK and Singapore, as well as US states California and Arizona, are establishing regulatory frameworks for them.

In addition, many companies have discovered that it's easy to replace all sorts of trips that used to take place by car with a two-wheel delivery or an electric van. As the service-to-the-home industry continues to expand, fewer city residents will feel the need to have a car. Fewer cars and less car use in cities means less gas guzzling. A lot less, in fact, given that cars account in some countries for up to 60 per cent of the oil consumed. The oil and gas industry will take a major hit.

Just as cholera was the impetus for the creation of Central Park, Prospect Park and Fort Greene Park in New York City in the nineteenth century, so will COVID-19 deliver more green spaces. Visits to urban parks and public gardens soared as much as seven times during the pandemic, once again demonstrating that public health is better served with green infrastructure, which includes bicycles.

More high-tech jobs are on their way too, with no small help from the bicycle. Electric bicycles in particular not only transform mobility; they also lead to the creation of high-quality jobs that promote digital and electrical skills in the workforce. With electric bikes spreading fast, these jobs are created on a daily basis, reinforcing the role of cycling in greening our economies and lifestyles.

Finally, we've run an intense global social experiment in working from home, and it appears to be a success, in part because the service-to-the-home industry stepped up. If only 10 per cent of us decide not to go back to a commercial office environment, the entire forward curve of commercial space would shift downward. That means the $2 trillion property investment industry is also in the process of being turned completely upside-down.

After coronavirus, people are much more conscious of the air they breathe and much less inclined to accept the enormous quantities of dirty air they were subjected to. If we reimagine the future by projecting forward from what two dozen cities have done with bicycles so far, we'll see that we've finally started effectively fighting back against the fact that 90 per cent of us worldwide have been breathing dirty air on a permanent basis. Because we can't see the pollution, we don't tend to think about it. But this poison weakens us all and kills nine million people a year, as well as placing an undue burden on health systems.

Why aren't the big polluted and congested cities in Latin America, Africa, the Middle East and South Asia deploying bicycles and electric bicycles aggressively? (India, for example, contains all ten of the top ten most polluted cities in the world.) All you need is bicycle lanes integrated, preferably, with the mass transit infrastructure. Make people feel safe riding bicycles and they will drive their cars up to 50 per cent less. Jakarta is a case in point. Famous for its congestion for decades (a colleague once spent seven hours getting to the airport from the city centre), Indonesia's capital and South East Asia's most populous city, with a population of 31 million, is rapidly transforming into an Asian (but regionally lonely) leader in mobility policy. Whipped into shape by an activist governor who is publicly leading the charge and regularly informing the public about new mobility policies, the city accelerated the roll-out of protected cycle lanes during the COVID-19 pandemic, all the while integrating these with mass transit options by providing bike storage and bike-share racks at subway stations, adapting subway cars to facilitate the transport of bicycles and increasing the total number of bike-share stations from 9 to more than 70 in just a few years. The result: cycling increased on a sustained basis, after the easing of social restrictions due to the pandemic, by 500 to 1,000 per cent depending on the area.

In many places, the car as a status symbol and excuses about the weather stand in the way of progress. However, these cities must wake up soon to the fact that an expensive, polluting car signals nothing more than disdain for fellow citizens, and that if the weather in Copenhagen or Amsterdam can be conquered by cyclists, so can the weather in Singapore or Cairo.

14

Fly Without Guilt

Flight shaming is out of control. People are increasingly scolded by others and criticized for flying, on the grounds that flying less mitigates the impact of emissions from aviation on climate change. The Swedish-born *flygskam* movement originated in 2018, when Swedish singer Staffan Lindberg pledged to give up flying. Sweden quickly saw a 5 per cent drop in the number of people flying (pre-pandemic), and that's incredibly rare: pandemic aside, the number of flights has been consistently increasing, to the extent that we have added 15 million flights per year in just the last 15 years to reach 40 million flights worldwide in 2019. *Flygskam* worked even better for Swedish domestic routes, where flights dropped 9 per cent. This doesn't make flight shaming right, however. Don't worry about it, fly without any guilt.

Flying less does of course decrease emissions: aircraft burn polluting jet fuel, and this in turn generates dangerous greenhouse gases that exacerbate the climate change catastrophe. Emissions from aviation are therefore a real and present danger: they increase each year and are expected to make up fully 25 per cent of our current global emissions by 2050.

The real issue, however, is not flying. It is instead the fact that governments subsidize the pollution from flying (jet fuel has been – scandalously – exempted from taxation for decades), that airline regulators ignore it and that progress towards the electrification of air transport has been held back. Airlines are allowed to use the skies as a public sewer at no cost, paying nothing for their emissions of harmful greenhouse gases. In addition, flying short haul would have been 100 per cent electric by now if we had not had to contend with 40 years of obfuscation and lies by Big Oil, which led to where we are today: cheap flights driven partly by the fact that Big Oil doesn't pay for its environmental destruction. Innovations for long-haul flights would be much further along as well.

When fighting climate change, we need to be courageous enough to acknowledge that we are fighting for system change, which is about addressing the root causes of the problem, instead of shifting responsibility to powerless citizens.

In 1982, ExxonMobil published a beautiful 46-page scientific report, with a 10-page detailed index of sources, estimating with stunning accuracy exactly the level at which emissions were going to be today. The scientists who wrote that report explained what that would mean in terms of climate impact: a catastrophe. ExxonMobil's science was spot on, and its scientific projections still hold up 40 years later. Then ExxonMobil lied, ExxonMobil lobbied and ExxonMobil profited, joined by all its industry peers. We now know, for example, that the American Petroleum Institute – the lobbying nerve centre of the US oil and gas industry – was publicly casting doubt on climate science and on the threat of climate change as early as 1980, and is still at it today, fighting against climate-friendly policies in at least 16 US states.

It's not that difficult for a trillion-dollar industry to obfuscate. The strategy is simple enough: throw a lot of money at sowing doubt about climate science by stressing uncertainty, while ensuring you don't employ any more pesky scientists unless they can be bought. To achieve that goal, Big Oil funded fake think tanks, fake scientists, bots and politicians to spread evident lies, falsehoods and lots of disinformation. Just the five largest publicly listed oil and gas majors, ExxonMobil, Shell, Chevron, BP and TotalEnergies, spend more than $200 million a year lobbying against climate science. That kind of money buys you a powerful voice.

In addition, the oil, gas and coal companies continue to seek to rig the system. Take the innocuously named Energy Charter Treaty, the little known, secretive and stealthy tool undermining climate action. The Energy Charter Treaty came into force after the end of the Cold War, ostensibly for a good cause: protecting energy investments by foreign investors in the former Soviet Union, before being widened to cover investments in any signatory country. It's a unique multilateral framework that more than 50 countries signed up for, including Russia, the European Union, the UK, Japan and several central Asian and Middle Eastern countries. Its sponsors continue to look to expand it to new states, particularly in Africa, Asia and Latin America.

But the Energy Charter Treaty is also somewhat peculiar. It has a 'sunset clause', which means countries can still face claims over decisions that affect their existing energy investments when policy changes – for example, when a country wishes to phase out coal-fired power – for 20 years. Its deliberations take place in private arbitration, outside national legal systems and their courts. Its cases aren't made public as plaintiffs aren't obliged to acknowledge the existence of a case, let alone reveal the compensation they are seeking; and finally, even if a state wins a case, taxpayers still have to pay its legal and arbitration fees.

It's a perfect venue for oil, gas and coal companies to derail climate action. A treaty is inflexible: unlike the law, it doesn't evolve with the times. When the US Supreme Court declared same-sex marriage legal in all 50 states in 2015, for example, it did so because cultural and political developments had allowed same-sex couples to lead more public lives, bringing about a shift in public attitudes, then questions about the legal treatment of gays and lesbians in courts. A treaty is different: it's self-contained and can't be changed as society changes. The Energy Charter Treaty has therefore become a perfect tool to seek compensation as societies have woken up to the climate catastrophe and tried to react.

The Netherlands, for example, moved to shut down coal-fired power plants because a Dutch court ordered them to do so. A German utility company, RWE, the largest polluter in Europe, sued the Netherlands, arguing that the Dutch government had invited them to build power plants and therefore they deserved to be paid $1.6 billion for closing them early. RWE used the Energy Charter Treaty to attempt to shift its losses onto Dutch citizens, seeking compensation for its own poor business decisions in building something that pollutes and kills. (RWE also owns dirty coal mines in Germany that have been the subject of long-standing protests because they are located in a 12,000-year-old forest, which the company is trying to destroy in order to dig for more coal. Despite this, it advertises itself as a green company, with taglines such as 'RWE: Welcome to the Renewable Age'.)

Two other companies, a German coal producer, Uniper, and a British oil and gas exploration company, Rockhopper, sued the Netherlands and Italy respectively on similar grounds. Uniper was objecting to the Netherlands ordering it to close a coal-fired power plant, while Rockhopper wanted to resist Italy banning new oil and gas drilling near its coasts because of the environmental destruction associated with that activity.

The Energy Charter Treaty was therefore turned from a tool to protect foreign investments to a major obstacle to averting an even worse climate crisis. Governments implementing measures to tackle the crisis are being threatened with enormous fines totalling tens of billions of dollars to save companies that took decisions they knew were harmful to public health and could worsen climate change. The abuse of the Energy Charter Treaty has reached such a dramatic stage that the European Union is now considering how it can either modernize it or exit it, while Italy has left it already and France and other countries are heading in that direction.

Yes, of course flying in jets powered by petroleum-based aviation fuel is terrible for the climate. Aviation (and its sister industry, shipping) is doing very little to reduce emissions and pollution from its activities. Legal constructs such as the Energy Charter Treaty stand in the way of progress by entrenching fossil fuel infrastructure and not allowing society the flexibility to move on. But the airlines and their regulators could do much more.

When grounded during the pandemic, some airlines – like Singapore Airlines and EVA – couldn't come up with anything more intelligent than to offer their customers flights to nowhere, attempting to sell seats on trips without a destination: planes took off, gave passengers a taste of their travel experience, then landed at the same airport. The lack of imagination of these airlines is stunning. They were encouraging carbon-intensive travel for no good reason whatsoever and passing the buck once again to customers, all to distract us from the fact that the onus should be squarely on them to pivot towards climate-friendly alternatives. The flights to nowhere are iconic examples of the deep-rooted structural problems of the airline industry and its unwillingness to do anything about them.

Another abusive carbon-intensive pandemic example came to light in Europe. Lufthansa, the German carrier, admitted in 2022 that it had been forced to run 21,000 empty flights that winter in order to meet a requirement by the European Commission that allowed it to keep its rights to take-off and land at European airports. Other European airlines were undoubtedly flying empty flights too, but Lufthansa was the only one to publicly disclose its data. Based on Greenpeace estimates, the number of these ghost flights aggregated across European airlines was at least 100,000. This was neither necessary nor wise in the midst of a climate emergency, and was caused once again by the fact that pollution from these flights isn't priced into their costs.

The global picture is even less encouraging. Back in 1997, the United Nations gave responsibility for international aviation and maritime emissions to the International Civil Aviation Organization in the case of airlines, and the International

Maritime Organization in the case of ships. Decades later, what have these two organizations done? Very, very close to nothing. They're still procrastinating. Even worse, several airlines are taking advantage of their customers by offering them the option to offset their travel through exorbitantly priced carbon credits that allow them to add to their profits while they continue to do little. As an aside, when you next see an airline offering to offset your travel by buying carbon credits, do not do so unless you know that you're buying them at the same price the airline is buying them at – and I can guarantee you that is not the case. But don't let that distract you from what matters. What matters is the electrification of air travel.

We lost 40 years of potential innovation in flight technology, battery research and hydrogen innovation because of the oil industry's obfuscations, and only now are we getting a move on. By 2022, finally unshackled from the fossil fuel industry's propaganda as net zero commitments started to gain momentum, 100 different electric plane projects were under development. Airbus, for example, released in 2020 details of three hybrid-hydrogen concept planes it said could fly by 2035, and by 2022 it had revealed that it was going ahead with testing specially adapted engines on an A380 superjumbo plane. In 2021, orders for electric aircraft started trickling in: DHL and UPS, the parcel delivery giants, announced that they would be buying 22 electric cargo planes for their package delivery businesses.

Furthermore, these 40 years also cost us in terms of slowing down innovation in rail travel, the best substitute for regional and short-haul flights. To paraphrase Thomas Edison, there's a way to do it better and we will find it.

We know that we're going to have many short-haul electric planes by 2030 because they already exist: for example, a passenger-grade electric plane completed its maiden flight in 2020, taking off from Cranfield Airport in England, powered by an engine that combines hydrogen and oxygen to produce electricity. British Airways is working with its manufacturer, ZeroAvia, as one element of its net-zero emissions strategy, and they are aiming by 2030 for an aircraft capable of flying 100 passengers more than 1,600 km. Trailblazing Denmark has already committed to having all its domestic flights free of fossil fuels by 2030. We also know that long-haul electric planes should be ready by 2040, thanks to announcements such as those by Airbus. Airlines that have the option to fly electric will do so, because the saving in costs is enormous: fuel costs savings alone are approximately 75 per cent.

Oil and gas companies need to compensate society for the 40 years lost to their disinformation. In the case of aviation, a surcharge should be levied on every airline ticket, paid for by Big Oil and directed to research into zero-carbon flights. There is a very good precedent for this. The Global Fund to Fight AIDS, Tuberculosis and Malaria was capitalized via a voluntary contribution from airline tickets. The fund to date has disbursed more than $20 billion and saved more than nine million lives (see more about this in Chapter 16, The Nasty Ninety).

For 40 years, Big Oil, Big Gas and Big Coal have got away with not applying the principle of 'the polluter pays'. For these companies, current and future damage from global warming are getting into the realm of the incalculable. The numbers are tremendous. We're talking anywhere from $400 billion to several trillion dollars depending on where we end up in terms of warming, because we need to take into account how climate change is destroying our planet's diversity, as well as the financial losses and human suffering that result from rising seas and temperatures. We shouldn't have to stop flying, nor feel guilty when we have to fly, because they lied for 40 years.

15

A Luxury Cruise Liner Is a Stinking Floating Dumpster

It's uncontroversial to suggest that when you break the speed limit, you should pay a fine, or that when you litter, you should pay a penalty. It makes sense, doesn't it? I say this because I don't understand why the same simple, logical, fair rules aren't also applied to corporations. Take luxury cruises. They involve giant boats that take their passengers to perfect vacation spots to unplug from the chaos of everyday life without having to wait in line at airports or drag bags from hotel to hotel.

I can understand why people might like luxury cruises. What I don't get is why 47 cruise ships owned by the world's largest cruise operator, Carnival Corporation, are permitted to emit 10 times more sulphur dioxide in European waters (2017 data) than all of Europe's 260 million cars combined. Do you know what price they pay for emitting all that poison? Nothing. Zero. Zilch. They are allowed to pollute free of charge. That's even though sulphur dioxide causes irritation of the nose, irritation of the throat, nausea, vomiting and stomach pain. It damages airways and lungs when you breathe it. It's just one of three air pollutants discharged from the ships' smokestacks, alongside particulate matter and nitrogen oxides.

Why is it that ships are allowed to pollute freely and not pay a penny for that pollution? Why don't we have stringent rules that stop them from doing it? Carnival Corporation owns 100 vessels across 10 brands, including Carnival Cruise Line, Holland America Line, Princess Cruises, Seabourn, P&O Cruises and Cunard. Since 2002, the company has been paying heavy fines for breaking the law, but they've paid far too few and only when they were caught red-handed. Polluting our lungs with sulphur dioxide is, amazingly, perfectly legal. They don't get fined for that. It's all the other stuff for which they've been penalized.

In 2016, Carnival paid $40 million – at that time the largest criminal penalty ever imposed for intentional pollution by a ship – for illegally dumping oil-contaminated waste into the sea and then trying to cover it up. In 2019, they were ordered to pay an additional $20 million for violating their probation terms; in other words, illegally dumping oil-contaminated waste into the sea and then covering it up, again.

The luxury cruise industry is a stinking floating dumpster in its entirety. Friends of the Earth ranks 18 major cruise ship companies based on four environmental criteria: sewage treatment, air pollution reduction, water quality compliance, and transparency. In its 2021 Cruise Ship Report Card, ten cruise lines received an 'F', while six received a 'D', one received a 'C' and one a 'B-'. Not surprisingly, all of Carnival Corporation's brands received an 'F', with Friends of the Earth noting that they had committed criminal environmental violations from 2017 to 2021.

Luxury cruise ships, a small segment of the shipping industry, are more visible than their commercial shipping brethren. We see them because they bring highly visible floating cities to dock near populated neighbourhoods and landmarks around the world. The pollution machines they represent, however, are replicated across the entire industry: shipping is likely the least regulated sector in the world in terms of emissions and one of the worst polluters across the board. It's somewhat shielded from view because its pollution is out of sight, far out at sea, out of mind and out of reach of any government, allowing it an easier path to cutting corners and saving money at every opportunity.

To start, the industry uses the dirtiest possible fuel to power its ships, one that is 100 times more polluting than road diesel. Bunker fuel is waste from the oil refining process and it is an environmental nightmare. Pitch black, thick, heavy and toxic, it doesn't evaporate and it emits more poison than other fuels because it's loaded with sulphur, which when burned releases gases harmful to human health and to the environment. These gases are poisonous to fish and aren't much good for seabirds or for humans living near ports, because the emissions travel hundreds of kilometres: approximately 70 per cent of shipping emissions occur within 400 kilometres of land. As a result, the industry emits a staggering amount of poison, much of which makes it into our lungs. Ship engine oil pollution increases incidences of childhood asthma and cancer, contributing to an estimated 1,300 premature deaths a year around the ports of Los Angeles and Long Beach and 60,000 a year globally.

By using this fuel, the industry is also subsidizing the oil and gas sector. If shipping lines weren't buying it, oil and gas companies would have to dispose of it safely, an expensive proposition. Estimates are hard to come by for the disposal cost saving, but this subsidy could be in the order of $450 billion.

The shipping industry dumps its plastic in the oceans. It dumps its trash. It dumps its excess oil. It also dumps its foul wastewater. In a wonderful illustration of the form-over-substance principle, ships are required to treat waste using equipment referred to as 'Marine Sanitation Devices' – in capital letters probably to make it sound better – except this can mean literally any equipment designed to treat sewage, often leaving behind toxic contaminants, bacteria and heavy metals that harm marine ecosystems.

The shipping industry is also responsible for over a billion tons of carbon a year. If it were a country, it would be the sixth largest polluter in the world. Its emissions are growing to such an extent that they are currently expected to be anywhere from 50 per cent to 250 per cent higher by 2050. It deploys more than 600 mega-containers, each of which can produce over its lifetime the same amount of pollution as 50 million cars. Put another way, this means that just 15 of these mega-container ships match the pollution emitted by *all the cars in the world*. Yet industry representatives such as the International Chamber of Shipping continue to argue that the sector is outside the scope of the Paris Agreement on limiting global emissions because it wants to remain self-regulated, code for doing very little.

At the same time, shipping is very important, because 90 per cent of global trade is carried via the world's oceans by 90,000 marine vessels. Every day, the stuff that supplies our supermarkets, car dealerships and shopping centres arrives mostly by ship: nine out of ten items are sent by sea.

The shipping industry's regulator, the International Maritime Organization (IMO), isn't doing anything much at all. It has not managed to negotiate a reduction in emissions and has no teeth to stop these 90,000 vessels from polluting at will and dumping their toxic garbage into the oceans. In 2020, hard on the heels of major oil spills in Mauritius, Venezuela and Russia and a supertanker exploding off the coast of Sri Lanka (to say nothing of the gigantic cargo explosion that flattened Beirut), the IMO even greenlit a package of fuel 'efficiency' measures that guarantees rising emissions until at least 2030, a stark reminder of its ability to operate with impunity.

It also continues to resist setting a net zero target by 2050, and postponed until 2023 efforts by some of its member states to force it do so. The UK, the US, France, Germany and others signed a declaration at the 2021 Glasgow climate talks to push for net zero on shipping emissions by 2050, while the European Union wants the industry to cut emissions by 55 per cent by 2030 and pay for the pollution generated by travelling to and from and within the EU by 2026.

The IMO's attitude cannot be right. If we individuals despoil the places where we live and work, we pay fines. Ships should be paying very large fines for what they're doing. Why is it that the citizen is always asked to be fair to other citizens, but corporations can do what they like?

Encouragingly, an astounding 40 per cent of everything we move by ship is oil, gas or coal, to be burned at destination to produce steel, cement, etc., or to provide us with electricity and heat. As we continue the inexorable move to a global energy system powered almost entirely by renewable energy, global shipping volumes should be cut by nearly half. With that, the pollution the industry is responsible for would be cut in half as well.

The waste built into the fossil fuel economy is a sight to behold. We must dig for coal, oil and gas, then build infrastructure to process it and move it around: roads, railways, an incredible seven million kilometres of pipelines, and ships. As soon as we've burned that coal, oil and gas, we have to go through the whole process again, ad infinitum. By contrast, with renewable energy, we simply need to find the material to build our solar panels and wind turbines, then ship these to where they need to be. Once we have done that, they can sit there for 20 to 30 years, powered by the sun and the wind. While we're waiting for oil and gas shipments to decrease, the shipping industry needs to be pushed to overhaul its practices across the board.

Every major shipping company should publicly commit to decreasing its greenhouse gas emissions to net zero by 2050 at the very latest, with a clear plan of how they intend to get there and milestone targets on the way: for example, a 40 per cent reduction within a decade. They could follow the lead of Maersk, the world's second-largest shipping line, which in 2022 accelerated its carbon neutrality goal by 10 years, moving it from 2050 to 2040 in a clear signal that zero-emission shipping is achievable. Furthermore, Maersk's carbon neutrality goal encompasses the energy consumption of its entire supply chain. Most importantly, perhaps, it is skipping a transition via natural gas to replace bunker oil as fuel for its ships (natural gas is awful – see Chapter 9 – but it is almost sulphur-free) and moving directly to low- and zero-carbon alternatives such as methanol. To that end, it is expecting the delivery of 12 methanol-powered cargo ships from 2024 and has been busy sourcing green methanol – derived from biomass or produced from green hydrogen and carbon dioxide – to power them.

To tackle air pollution in our cities, ships docking at port should be required to use electricity from onshore sources at all times, and pay for it. Vessels without this capability should be retrofitted at their owners' cost. It's incredible that in 2022 they aren't required to use onshore electricity, and often can't, because the infrastructure is lacking either on the ship or at the port; in 2022, half of Europe's ports still lacked onshore power supply infrastructure, as did over 85 per cent of US ports. The overwhelming majority of ships therefore keep their auxiliary diesel engines running, spewing pollution freely (including particulate matter, nitrogen oxides, ozone and air toxics). They're happy because they save money – marine fuel costs are cheaper than electricity because pollution isn't accounted for either in the cost of the fuel or in the price society pays for the emissions – but an IMO- or government-led initiative to require onshore power would lead to zero emissions and pollution from ships docking at ports around the world, saving countless lives.

A lot more can and should be done, including an end to water pollution (if it were mandated, for example, that sewage systems should be upgraded from mysterious Marine Sanitation Devices to advanced wastewater treatment systems) and compulsory public reporting by shipping companies of their air emissions and water discharges. Meanwhile, we need to recognize that the system is rigged in favour of shipping companies in ways that cost us our health and our environment, and that must change. We can start by skipping the well-earned luxury cruise: it's just a floating garbage trash can.

16

The Nasty Ninety

The record of global corporations speaks for itself. Just 90 companies headquartered in 43 countries are responsible for two thirds of all the bad greenhouse gases that we've emitted since industrialization began. *All* of these 90 companies are oil, gas, coal or cement companies. In other words, these 90 companies have been frying the planet pretty much on their own. Fifty of them are investor-owned companies listed on stock exchanges. Thirty-one are state-owned companies such as Saudi Aramco. Nine are government-run industries in countries such as China, Poland and Russia. Fifty-six are oil and gas companies. Thirty-seven are coal producers (including subsidiaries of the oil and gas companies). Seven are cement manufacturers.

More broadly, analysis of 6,000 listed companies routinely shows that they aren't doing enough (or often anything at all) about global warming, despite the profusion of announcements that many are aiming to become carbon neutral. Instead of working to meet the targets of the 2015 Paris Agreement to tackle climate change, which aims to limit global warming to well below 2° Celsius, the actions of businesses globally are on course to deliver a 4° rise.

We keep reading about how worried businesses are about climate change, or about how many of the world's biggest companies are braced for the prospect that a warming planet could substantially affect their bottom lines within the next five years. For example, an international not-for-profit called CDP (formerly the Carbon Disclosure Project) calculates that hundreds of companies potentially face $1 trillion in costs related to climate change in the decades ahead. Fine. But we should focus instead on the environmental destruction that these corporations are causing now, which is exponentially more important. If they stopped what they are doing and started acting responsibly, we would all be healthier and wealthier.

I'll come back to corporations in a moment. But first let's take a detour into the fight waged in the 1980s and 1990s against AIDS. In that fight, a generation of campaigners coordinated global action against powerful companies that were denying the AIDS problem and vested interests that were shamelessly financing armies of lobbyists and lawyers worldwide in order to protect themselves to the detriment of those with HIV. Sound familiar?

Campaigners against HIV denial were battling first and foremost the world's largest publicly traded pharmaceutical companies, because Big Pharma was then producing HIV medication that only the super-rich could afford rather than drugs that could be distributed widely to treat or prevent the disease. In a fight that lasted 20 years, the campaign managed to transform access to this treatment into a human right for all, a generic medicine funded by an international financing organization.

AIDS was first diagnosed in 1981. At the time, it was widely misunderstood and misrepresented by powerful voices in politics and in society. These voices sought to minimize the threat it posed and to marginalize and ostracize the victims. But the suffering caused by HIV/AIDS was enormous and widespread. A social movement arose demanding universal access to medication for people living with the disease. It finally succeeded in 2001, when the UN General Assembly declared that everyone had a right to treatment.

By 2008, more than four million people in poor countries were benefiting from what had hitherto been unaffordable drugs. No one should diminish the struggles some HIV/AIDS sufferers still face today getting access to medicine. But we do have to respect the astonishing success of that generation of HIV campaigners. The bungling, stumbling, inefficient climate movement needs to learn from its example.

The HIV movement succeeded because of five factors.

First, it crystallized its struggle by focusing just on Big Pharma. HIV campaigners did not target other corporations, hospitals, politicians or newspapers, even though each was implicated in what was happening. It found and zoomed in on Big Pharma's weak points and attacked it mercilessly for favouring profit over people. The climate movement, on the other hand, is bumbling and stumbling because it has far too many targets and is fighting too many battles at the same time.

Second, HIV campaigners had a simple compelling message: that the drugs necessary to fight HIV were essential for life.

Third, the movement was coherent: it focused on clear goals based around universal access to treatment, without regard for a sufferer's ability to pay.

Fourth, it convinced the public and politicians in multiple countries that the cost of effecting change was relatively low because all they had to do was manufacture cheap generic drugs.

Finally, it succeeded in pushing the world to create stable institutions that gave the movement permanence. The main institution is the Global Fund to Fight AIDS, Tuberculosis and Malaria, which to date has saved over 38 million lives and spends $4 billion each year to accelerate the end of the three diseases.

If we are to apply the lessons from the fight against AIDS to the struggle over climate change in order to maximize the chances of success, we must first narrowly define the targets of climate action. We know that emissions come from many activities. We also know that emissions come from many energy sources, like oil, gas and coal. These multiple targets contribute to the huge size of the climate movement, but also to its propensity to expend its energies in many directions at the same time. We're not going to get anywhere by trying to cover all the bases, because in reality the problem can be very specifically defined.

First and foremost, we have to focus on the 90 companies responsible for two thirds of the harmful emissions generated since the industrial age began. These 90 companies control five times as much oil, coal and gas as it's safe to burn today, and 80 per cent of their reserves must be locked away underground to avoid the coming catastrophe. This tiny collection of large companies lobbies hard to prevent climate action, not least by spending billions to confuse and distract us. But they're at the centre of the model of

146

intensive carbon use that's destroying the planet. Just like the HIV movement, climate campaigners have a relatively small target on which to focus their energies.

As the HIV movement showed, we also need a compelling message and framework. Applying this lesson, we must focus on *people*, and specifically on our right to life, i.e. the human rights violations of these 90 companies. Climate action would be far more effective if it concentrated on human suffering in a way that was communicated consistently and clearly to everybody.

We also need to be coherent. We know that we need a drastic cut in emissions. We know that we need an economy fuelled by renewables. The challenge is that the cost of climate action is not always acceptable because the damage caused by fossil fuels is not properly understood. The largest 3,000 companies in the world by value cause $2 trillion of environmental damage every year. The global climate action we require would cost a fraction of this amount.

Keep in mind too that when human rights are being violated, costs must take a back seat. For example, ending the Atlantic slave trade was extremely costly to slave owners and to the UK exchequer, which had to compensate them. But who cares? These costs were immaterial because slavery had to be stopped. Climate campaigners simply have to focus like laser beams on the main culprits — these 90 companies — and pursue far narrower objectives. Our goal should be to shrink the market capitalization of these bad actors by making their money much more expensive. They still have access to hundreds of billions of dollars of very cheap capital, in the form of either investment or borrowing, and that needs to change

Fourth, the price of change has to be low. In the context of the climate fight, this isn't very difficult. The best way would be to compel the capital markets to increase the cost of money for these 90 companies. The world's capital markets are a $230 trillion monster: this represents all the money in the world available to fund companies. Structurally, this money comes ultimately from the public and governments, and then flows to pension funds and endowments, which sit at the apex, before reaching asset managers, banks and ultimately businesses and other borrowers.

Pension funds represent 25 per cent of the world's capital markets. In the United States, just 75 funds account for 80 per cent of the money in the industry. These 75 are, when they want to be, influencers. When they move, everyone moves. All these pension funds have trustees, human beings who owe us a duty to invest prudently, but they aren't discharging that duty: although we know that 80 per cent of oil, gas and coal reserves cannot be extracted without extremely serious consequences for our planet, pension funds, stock markets and bond markets continue to assign value to the 90 Big Polluters based on these future supplies.

The trustees of the pension funds that sit at the top of the capital markets control how money flows and on what terms. It is high time they realistically factored climate-related risks into their investment analysis. If they did, large amounts of money would migrate from the bad guys to the clean energy economy. The value of the 90 companies would collapse, and they would no longer be able to burden us with the reserves that they want to extract and burn irrespective of the cost to the planet.

One way of bringing pressure to bear on pension fund trustees is to pursue them through the legal system, via targeted lawsuits wherever legal analysis shows that litigation has a chance of succeeding. If those lawsuits were unleashed en masse, the pension fund industry would change the mandate it gives asset managers. Money would then flow in the right direction.

The fifth and final lesson from the HIV fight is the need for stabilizing institutions. In the case of the climate fight, these institutions are already around us. There's a huge United Nations climate finance infrastructure that needs to be strengthened and revived. The UN needs to streamline what it's doing, and we need new mechanisms to fund it, such as voluntary contributions by airline passengers or cruise ships, in a manner similar to the way in which the Global Fund to Fight AIDS, Tuberculosis and Malaria was launched. Or perhaps there could be a tiny levy on share, bond and derivative transactions carried out by banks and hedge funds. Just 0.01 cent on every transaction could raise hundreds of billions of dollars every year. We would then have institutions with the financial muscle to help the vulnerable while the capital markets shift their cash to fuel clean lifestyles and economies and stop funding oil and gas companies – particularly the Nasty Ninety.

We can fight climate change effectively and coherently by targeting hypocritical corporations, without having to worry about fighting governments and consumers at the same time. We have a defined target of just 90 companies. Changing their behaviour *will* change the world. We can produce a compelling message that aligns with human dignity and rights. It isn't expensive to apply pressure to pension fund trustees – the flurry of lawsuits mentioned above would do it – nor would it be terribly hard to strengthen our institutions: governments are already aligned, via the Paris Agreement, on the need for decisive climate action.

We need to focus our energies on the Nasty Ninety. Their significance dwarfs all other green issues. If the Nasty Ninety corporations change, everything else will follow.

17

The Social Media Axis of Evil

You may have heard, seen or read that Big Tech – companies like Google, Facebook, Apple, Amazon and Microsoft – have all decided they want eventually to be 100 per cent powered by renewable energy. We the customers want it, and therefore they have been working on green-upstaging each other for a few years now, with Google most recently taking the lead by pledging that it will run its operations purely on carbon-free energy by 2030 (Amazon promises to do the same thing, but by 2040). To sell this and improve its environmental credentials, Google is assuring us that every email we send through Gmail, every question we ask it, every YouTube video we watch and every route we take using Google Maps is a service powered by renewable energy.

In some ways, these green pledges are great, because every step large corporations take to clean up their act helps create new norms. But while it's nice to know that every search powered by Google is delivered to us using green electrons, it also matters what the company (and the others) are using their tremendous artificial intelligence, big data and emerging robotics skills for.

While advertising green credentials, Big Tech is also, covertly for the most part, forging lucrative global partnerships with oil, gas and coal companies, with the sole purpose of increasing fossil fuel production. That act alone places them squarely in the Axis of Evil category. We already know that despite the global climate emergency we're in, Big Oil is doubling down on fossil fuels. Companies like ExxonMobil, Shell, TotalEnergies and Saudi Aramco are scrambling to produce more of them than ever before. They're getting plenty of help from banks, and from governments around the world too.

What we don't pay much attention to is the fact that Big Tech is building a new and overwhelmingly harmful carbon cloud, by placing their computer power at the service of oil companies so that the latter can use their cutting-edge technologies to produce even more oil, faster and cheaper.

You would think that technology companies are leaders in corporate sustainability, given how public they are with their save-the-world rhetoric. However, in 2019, an article penned by an anonymous Microsoft software engineer was among the first to contribute to exploding that narrative. He was sent to Kazakhstan to work with Chevron, the US oil corporation, and the Kazakh state oil company to improve their extraction output with machine-learning techniques, big data and artificial intelligence – basically all the computer power Microsoft could muster. Chevron had signed a seven-year deal worth hundreds of millions of dollars to establish Microsoft as its primary cloud provider.

We need to digress for a moment to talk about cloud computing. It's basically a way for companies to rent computer servers instead of buying them. A few years ago, a company would have run its website from a server that it paid for and maintained itself. Today, the same company can outsource its infrastructure needs to a cloud provider such as Amazon, Google or Microsoft. The market is dominated by Amazon, and its cloud business now makes up more than half of all its operating income. In 2014, its cloud revenues were $4.6 billion. Just five years later, those revenues had rocketed to $36 billion.

Companies like Chevron are the perfect customers for cloud providers because for years they've been generating enormous amounts of data from their oil wells. Chevron alone has thousands of such wells around the world, and each is covered with sensors that send data to the company's servers or the cloud, generating a vast 1,000 gigabytes per day. Its computational needs are huge. But not always. When the analysis is complete, the company's computational requirements go down dramatically, and these sharp fluctuations mean that Chevron is not best equipped to deal with its own data ocean.

That's where the public cloud comes in: oil companies can solve the computer challenge by renting as many cloud-based servers as they need, whenever they need them, and paying only for what they use. But Big Tech doesn't just give us the cloud. It also provides analytical tools like artificial intelligence and machine learning to study the data.

Usually hydrocarbon companies find oil and gas by performing a seismic survey. They send sound waves into the earth and then analyse the time it takes for those waves to be bounced back from various geological features. The data generated is staggering – it can run to more than a million gigabytes – and the output is a

3D geological map, if you can imagine that for a moment, that geophysicists then study to recommend the best locations to build wells. It takes them months sometimes to interpret these maps, and it's hugely labour-intensive.

At least it was until recently. Computer visioning technology can automatically segment different geological features to help the geophysicists understand the 3D data and identify where best to drill. It places the computing power of Big Tech at the service of Big Oil to advance Big Oil's core priority: to dig out more fossil fuels from the ground while simultaneously cutting costs. The climate emergency be damned.

Let's get back to that Microsoft engineer. In recent years, Big Tech has aggressively marketed the transformative potential of the public cloud to Big Oil. In 2017, Microsoft signed the seven-year contract with Chevron that I just described. In 2018, it announced major partnerships with BP and Equinor. In 2019, it signed a deal with ExxonMobil, which ExxonMobil claims is the largest contract in cloud computing for the Big Oil industry.

Amazon is also in on it. Not content with the fact that up to 465 million pounds of the company's plastic packaging waste, all of it made from oil, already pollutes the world's oceans (about the same as loading a delivery van with plastic and dumping its content in the ocean every hour), they have also opened an office in Houston, the US oil and gas capital, to focus on servicing oil and gas clients. Google too has developed deep relationships in the industry, and has TotalEnergies and other oil companies as clients. Whatever Big Tech is telling Big Oil, it seems to be working.

Microsoft sent their engineer to Kazakhstan to integrate their AI and machine learning at the massive Tengiz oil field. According to the engineer's account, his counterparts at Chevron were tasked with boosting oil production by 50 per cent to one million barrels a day. This is at a time when we can't afford to extract a single new barrel of oil from the ground. If we're to stabilize climate change damage, we have to leave all of the world's fossil fuel reserves in the ground, peaceful, untouched and unexploited.

In other words, Microsoft and Chevron are going about their business maximizing oil and gas production without any regard for the climate emergency. In that context, the promises of big technology companies to be powered by renewable energy mean little except as examples of greenwashing.

While the Microsoft engineer was there, the Chevron managers – all American men, with not a single Kazakhstani in the room – wanted to talk about another idea. There were a lot of workers in the oil fields, and they thought it would be useful to know where they were and what they were doing at all times. Up to 40,000 employees were on site, nearly all low-paid Kazakhstanis, who worked rotating 12-hour shifts for two weeks at a time to keep the oil field running 24/7. The Chevron managers wanted to use Microsoft tech to keep a closer eye on them, analysing video streams from CCTV cameras and data from GPS trackers to monitor everything they did, right down to how frequently they took bathroom breaks.

Big Tech has invested billions in cloud computing infrastructure, and it clearly thinks that there's a critical path to recovering this investment and making even more money by capturing the technology spend of Big Oil. It's therefore doing everything it can to strengthen the revenues of the oil and gas industry. In emerging markets like Kazakhstan, we also know, thanks to the anonymous Microsoft engineer, that Big Tech is delivering the full dystopian package, right down to constant worker surveillance.

Then there is Autodesk, not quite a household name but nonetheless a multibillion-dollar software firm that employs more than 10,000 people and plays a pivotal role in the architecture, engineering and construction industries. The California Big Tech company 'makes software for people who make things', and its products make it easier to design buildings and structures – for oil, gas and coal extraction, for example. Autodesk is possibly Big Tech's loudest climate advocate, marketing itself for the whole of the last decade as a company committed to environmental sustainability, loudly proclaiming that it views corporate responsibility as a core business value. But it doesn't practise what it preaches. Rather, it's greenwashing its reputation while enthusiastically turning a blind eye to how its software products are used.

Like other Big Tech companies, Autodesk has a 'denial list' of off-limits clients (these typically include manufacturers of weapons of mass destruction, for example, or terrorist organizations). Perhaps it's time for these denial lists to include fossil fuel companies. Humanity can't afford any new oil, gas and coal extraction if we're to ensure that we don't overheat the earth beyond its ability to host us. We consume about 100 million barrels of oil a day, but we need to bring this down to 10 million by 2050 to avoid triggering a cascading set of tipping points that lead the world into a hothouse climate state.

In one egregious use of Autodesk's world-leading software, the German energy giant RWE is destroying part of what remains of Hambach Forest, an ancient woodland near Cologne, to extract lignite, a sedimentary rock formed from naturally compressed peat. It's the dirtiest form of coal, the one most harmful to human health. RWE is using 90-metre-tall diggers to extract this poisonous coal, and these excavators wouldn't last long without Autodesk's 3D-design software. Put another way, without Autodesk, no new lignite could be extracted and the destruction of an ancient forest would cease, an objective local communities and climate activists have been relentlessly pursuing for more than 40 years. It's highly unlikely that Autodesk haven't noticed their products being put to this use, or that their decision to keep supplying RWE with critical software is innocent.

Big Tech is not responsible for Kazakhstan's reliance on oil or RWE's destruction of ancient forests, and it isn't to blame for the climate emergency. It is, however, exacerbating climate change through its partnerships with Big Oil and energy companies intent on extracting dirty coal at all costs. Keep in mind that climate change is a problem where the damage done by the relatively few (mostly oil and gas companies) dwarfs the attempts to make it better by the very many (consumers mindful about single-use plastic, or planting trees, for example). Big Tech has a choice: it can empower one side or the other. Right now, it's clear which side it's really on, and it's not the one where we find our children or grandchildren.

This simply has to change. The simplest, most straightforward way to starve Big Oil and companies like RWE is to increase the cost of their capital – through a credible carbon price, for example – and make their money a lot more expensive. In this toxic relationship between Big Tech and the fossil fuel sector, we need cloud and software providers to simply refuse to service oil companies. It's an approach that would mirror that of the fossil fuel divestment movement, which in just a decade has caused institutional investors to sell oil, gas and coal stocks en masse, as well as successfully question whether Big Oil are still entitled to their social licence to operate.

While fossil fuel divestment initiatives existed before, the movement achieved scale in 2012, when environmentalist and campaigner Bill McKibben published an article in *Rolling Stone* magazine, founded advocacy organization 350.org and launched the Go Fossil Free: Divest from Fossil Fuels! campaign. Activist college students, first in the US then in many other countries, suddenly had a coordinating body to turn to for support and to broadcast their calls for universities to sell their oil, gas and coal stocks. In less than a decade, the movement went global and broadened to include faith organizations, institutional investors and local government pensions. Irrespective of whether selling stock in an oil company affects its strategy (though buying stock and taking control would), the divestment movement socialized the fact that fossil fuel companies are intent on burning all identified reserves as well as any new ones they find, regardless of the effect on the climate, and that this needed to change. In addition, it has contributed to weakening their social licence and given momentum to a rapidly rising wave of climate lawsuits around the world.

In the same way, NGOs and charities could focus on pressurizing Big Tech to simply refuse to host oil companies in their cloud computing infrastructure or sell them critical software they need to extract more oil, gas or coal. Big Tech has had an easy ride on climate action. It's time to hold it to account.

18

Dial Down That Air Conditioning, But Not Too Much

On 17 May 1906, Stuart Cramer, an American engineer and inventor, gave a speech to the American Cotton Manufacturers Association's 10th annual conference in North Carolina. It was entitled 'Recent Development in Air Conditioning'. Little did the US Naval Academy engineering graduate know that he had inadvertently named an invention that was going to change the world.

Air conditioning is the process of removing heat and controlling the humidity of air. Cramer wasn't its inventor, though. The introduction of modern electrical air conditioning is credited to Willis Haviland Carrier, an American inventor who installed the first unit in 1902 at a publishing company in Brooklyn, New York. His eponymous Carrier Global Corporation is to this day one of the world's leading makers of air conditioning and cooling systems.

The rise of air conditioning was slow at first, but took off dramatically after World War II. Air-conditioning units were adopted by cinema theatres in the 1920s, but were too large and expensive for installation in homes until 1947, when their newly compact size and affordable cost allowed their rapid deployment to begin.

Advertising in the 1950s focused on American women at home, extolling the advantages of air conditioners as appliances they should be proud to own because of their outstanding design, their availability in each season's smartest colours and their contribution to better living, better health and better comfort. Smiling women were sometimes accompanied by sweating men returning home after work, about to bask in the comfort of the cool air inside.

Today, over 90 per cent of American and Japanese households have air conditioning, as do 60 per cent of Chinese homes, though penetration in India and Brazil lags at 5 per cent and 16 per cent respectively. Its transformational power has been immense. Singapore's founding father, Lee Kuan Yew, liked to say that air conditioning made development possible in the tropics, and its installation in buildings where the civil service worked was the key to the public efficiency of the city-state, crediting the invention with nothing less than the rise of Singapore.

In the US, the number of people living in Florida, southern California, Texas, Arizona, Georgia and New Mexico exploded after World War II, in no small part because of air conditioning. Dubai probably wouldn't have become a global financial centre without it. The information technology industry also owes it a debt of gratitude: The cooling technology pioneered by air conditioning allowed the world to be blanketed with computers despite the enormous amount of heat they generate from the electricity at work inside them.

Crucially, air conditioning and related cooling technologies also save lives by cutting exposure to sometimes unbearable heat, and enhance our quality of life in so many ways, such as protecting perishable food and, as demonstrated in the COVID-19 pandemic, allowing new technologies to develop for vaccines that require freezing. Air conditioning doesn't make the news headlines often, but its profile has recently risen. It's a hot global-warming topic – literally and figuratively – because systems that keep us cool are among the biggest guzzlers of energy in the world.

As the planet gets hotter while the global population increases and continues to urbanize, the number of air-conditioning units is set to triple by 2050 to six billion. India alone is expected to install 1.1 billion units by 2050. Energy demand from cooling is on track to reach 13 per cent of electricity demand by 2050. According to the US Environmental Protection Agency, heatwaves in the continental United States are occurring three times more often than in the 1960s, and partly as a result, the amount of energy use in the summer has doubled since 1973, with air conditioning alone accounting for 17 per cent of the average American household's energy consumption.

We can't ignore air conditioning as we decarbonize. Ownership of air conditioners increases with rising incomes as well as with a warming world, yet at the same time, they produce enormous heat themselves and contribute to higher urban temperatures while leaking potent greenhouse gases and putting pressure on 'peak electricity' (demand is most acute during the day, when overall need for electricity is already high).

Air conditioners contain chemical blends that cycle through them, called refrigerants, which absorb heat from indoor air and transform it into cold air. This blows out via a fan, while the warm refrigerant is sent outside. These refrigerants have had a turbulent history. At first, they were often toxic, flammable or both. Leaks required rapid evacuation of factories and other spaces being cooled. Then, in 1928, refrigerants called CFCs (short for chlorofluorocarbons) were invented. They had the huge advantage of being safe: they were non-flammable, non-explosive, non-corrosive, low in toxicity and odourless. They were widely used for decades, until it was discovered in 1974 – a discovery that subsequently earned the scientists involved, Paul J. Crutzen, Mario J. Molina and F. Sherwood Rowland, the Nobel Prize in Chemistry – that they were depleting the ozone layer, a thin part of the earth's atmosphere that absorbs almost all of the sun's harmful ultraviolet light. In other words, while helping us stay cool, CFCs were slowly killing humanity.

Big climate problems require massive cooperation between all countries, and rapid solutions. The world delivered on CFCs, and in 1987 a treaty known as the Montreal Protocol was signed and ratified in record time to stop the production of CFCs and other ozone-depleting substances.

However, we needed replacements for CFCs in order to stay cool, so in came HFCs (short for hydrofluorocarbons), another greenhouse gas but this time with zero ozone-depletion potential. Soon enough, though, we woke up to the fact that the first HFCs (there are many types) had a global-warming potential of 1,400 relative to carbon dioxide. That's better than CFCs' potential of 5,000–12,000, but still a climate change scourge.

Today the world is busy developing HFCs with low global-warming potential, but it's also using a newly found hammer:

mandates by cities requiring buildings to reduce their overall emissions. The trailblazer was New York City, which passed legislation requiring all large buildings to reduce their overall emissions by 40 per cent by 2030 and by 80 per cent by 2050. Efforts are under way everywhere to match New York's muscle, some of which are predicated on advances in refrigerants that have a low global-warming potential, as well as on improvements in the basic technology: the best available air conditioners are already twice as efficient as the average air conditioner on the market.

Conventional wisdom holds that even if all air-conditioning units worldwide become more efficient, we are going to have so many of these that they will put enormous pressure on electricity supply. Implicit in that wisdom embedded in projections and publications by the World Economic Forum, the Lawrence Berkeley National Laboratory, the International Energy Agency, the Rocky Mountain Institute and others is the fact that it's not okay for Indians, Africans and others from developing nations living in hot climates that are getting hotter to have access to air conditioning on the scale that Americans and Japanese do.

The scaremongering is often based on projections of massively large climate-warming emissions from air conditioning that will account for 20–40 per cent of the world's remaining carbon budget. The carbon budget is a metric measuring how many more carbon emissions we can dump into the atmosphere before global warming exceeds 2°C above pre-industrial levels – the goal set at the Paris climate conference in 2015.

It's true that perpetually cooling the vast volumes of hot air that fill homes, offices and factories will always be a massive guzzler of energy. The problem isn't just that more air conditioners will require ever more electricity to power them. It's also that they'll particularly boost the amount that's needed during peak times, when temperatures are really roasting and everyone's cranking up their AC at the same time. That means we need to overbuild electricity systems to meet levels of demand that may occur only for a few hours of a few days a year.

But fears of more air conditioners are overblown. Mandates by cities requiring buildings to reduce emissions are the over-arching tool we need everywhere: this will accelerate the adoption of increasingly efficient air-con technologies and of smart grids – sensors, control systems and software – that automatically reduce usage when outdoor temperatures decline or when air-conditioned spaces are empty. We can also cut direct emissions from air conditioning by switching to alternative refrigerants that exist already. The world has already agreed via a 2016 amendment to the Montreal Protocol to shift to refrigerant options with lower warming impacts.

Simple energy conservation measures increasingly adopted in many cities will also help: adding insulation, creating more green spaces inside built environments, covering windows and installing cool roofs designed to reflect more sunlight and absorb less heat than a standard roof. So will government mandates such as that in Japan recommending that thermostats are set to a quite hot 28° Celsius in both government and private offices. Japan's 'Cool Biz' campaign, in place since 2005, encourages casual wear in summer as well as moderate use of air conditioning, and has been actively implemented across the country. India has taken a different approach, mandating that all new air-conditioning units sold in the country have a default setting of 24° Celsius, counting on the fact that most people won't change the default setting once they turn their machines on.

The most crucial fix, however, will occur outside the air-conditioning industry. We are moving inexorably to a world powered by clean and renewable energy. Even that historic bastion of fossil fuel interests, the International Energy Agency, has revised its projections upwards for solar and wind power and now admits that not only did global renewable energy deployment grow at the fastest pace in two decades in 2020, despite supply chain disruptions and construction delays due to the impact of COVID-19, but that going forward, renewable capacity will account for 90 per cent of the entire global power sector's expansion.

Adoption of renewable energy is growing exponentially, and at current rates this deep transformation of the energy sector globally will deliver quasi-total clean energy everywhere by 2040 or before. By definition, this will give us an overabundance of power: we need more solar, wind and battery power than the electricity we consume in a steady state in order to ensure uninterrupted supply.

Singapore, for example, has announced that it will move its 95 per cent natural-gas-dependent electricity system to, initially, 30 per cent renewable energy by 2035 by importing it from neighbouring countries. This requires it to add approximately four gigawatts of electricity generated from new solar power plants located elsewhere, as it retires some of its old natural-gas-fired power plants (four gigawatts is enough to power at least three million homes). However, Singapore also wants that solar power to be delivered to its shores at a 75 per cent capacity factor. Because the sun doesn't shine all the time, a solar power plant will habitually operate at a capacity factor – the ratio of electricity the plant produces over its total potential if it was running all the time – of anywhere from 10 per cent in a cold and cloudy location to just over 30 per cent in a sunny location. As a result, to ensure that four gigawatts of solar power can be delivered to Singapore at a 75 per cent capacity factor, current engineering requires solar power plants capable of producing approximately 20 gigawatts to be built, supplemented by battery storage. Not all of the 20 gigawatts will be required because sun cover will vary daily, and this capacity would be built under the worst-case solar irradiation scenario.

This is an example of the overabundance of energy that is a feature of electricity systems powered by renewable energy and that will power zero-emissions air conditioners wherever these are required, steadily reducing their related direct and indirect greenhouse-gas emissions.

19

Going Vegan to Go Green? Don't Bother

There's been a tsunami of exciting new arrivals on the global food scene, many mimicking meat by ingeniously using and mixing common vegetables, vegetable oils and legumes such as soybean. When I tried for the very first time the wonderfully named Impossible Burger, this 100 per cent plant-based burger with zero beef tasted just like a normal burger. Frankly, it tasted like one of the best gourmet ones rather than the cardboard burgers we might expect from a well-known fast food joint.

The Impossible Burger and its cousin Beyond Burger, both tasty faux-meat burgers, were created by vegans. Their widespread adoption was subsequently spurred by vegetarians, who historically have been (and still are) more numerous than vegans. These burger lookalikes are an excellent example of how behavioural change can be a springboard for new technologies. Not only are they now sold in Burger King and McDonald's, but the combined market value of Impossible Foods Inc. and Beyond Meat Inc. is over $10 billion – all of which is wealth created from thin air in under 10 years.

Powering their popularity is the rise of veganism: British vegans numbered approximately 1.5 million in 2020, up 40 per cent in one year, while American vegans increased six-fold between 2014 and 2021 to 6 per cent of consumers (the number of vegetarians, by contrast, has changed little in recent years).

Traditionally veganism was motivated by a central ethos of doing no harm. As more people became aware of the cruelty to which animals are subjected on their way to the dinner table, the number of vegans rose. But then another key rationale evolved: the public learned of the reckless environmental practices of the meat industry via documentaries such as *Cowspiracy* in 2014, strengthening the veganism movement as it became widely known that the carbon footprint of meat – especially beef – tends to vastly exceed its calorific equivalent in vegetables. Personal health concerns also play a role for some, who adopt plant-based diets to avoid processed foods (even if the latter are also prevalent in vegan diets).

I loved my Impossible Burger, but I won't join in the flagellation to which many vegans feel they have to subject themselves, and others, about beef.

I've been at meetings, lunches, dinners and Zooms where someone plays the vegan card like an ace in the hole. By all means don't eat beef if you don't like it, if you believe it harms your well-being or if you're concerned about animal welfare, but it's important to recognize that it isn't going to do much to save the planet from climate change: eating less meat may be a good thing, but its impact is limited. After all, the world continues to eat more meat each year than ever before because its consumption is associated with higher living standards, and the fastest growth is driven by increasing wealth in low- and middle-income countries.

Beef in any case is not the problem, its pricing is: beef sells at a price that doesn't reflect its environmental footprint and is therefore cheaper than it should be. Exploiting the trend to demonize it gets us nowhere. Epicurious, the American digital brand that focuses on popular recipes for home cooking, has stopped publishing new beef recipes to apparently encourage more sustainable cooking. Such a policy is misdirected and uninformed: I don't know anyone who needs a recipe to barbecue a sirloin steak.

Vegetarianism pre-dates the modern era and the climate crisis. Its first philosopher-proponent was Pythagoras of Samos, 2,500 years ago. It has been with us ever since, based on the concept of benevolence towards all species and not causing pain to other animals. However, it isn't a magic bullet that will solve the beef problem, because we won't all become vegetarian. We *are* solving the beef problem (more on this later), and those who don't want to go vegetarian or vegan don't need to.

The real problem remains the same – burning gas, oil and coal – and that's what we have to stay focused on.

Anyone who uses the words 'environment' or 'climate change' to contextualize their vegetarianism or veganism should be aware that they may be under the influence of the public relations departments of oil, gas and coal companies, who have been successfully distracting us for decades. As discussed elsewhere, they have implemented a strategy to shift responsibility for the climate catastrophe by encouraging and embodying narratives fixated on individual responsibility in order to deflect a collective push for what we really need: systemic change such as taxing carbon, decommissioning fossil fuel infrastructure and banning new oil and gas exploration. As a result, citizens concerned about climate change might feel they need to focus on going zero beef or zero waste, a narrative that is highly effective because it capitalizes on our sense of responsibility while fooling us with easy solutions: eat less beef, fly less and recycle, and all will be fine.

Then there are the deceptive climate ads, thousands of them over decades, with headlines such as 'Doomsday is cancelled' or 'Who told you the earth was warming?', or advertorials seeking to downplay risk in the run-up to climate negotiations, such as ExxonMobil's advice ahead of the Kyoto climate talks in 1997 to 'not rush to a decision at Kyoto … We still don't know what role man-made greenhouse gases might play in warming the planet.'

We need to be aware of this and not let them get away with it. So how dangerous is beef really, and do you have a moral duty to avoid it?

There's a very bad greenhouse gas called methane that's substantially worse than CO_2 in its warming impact. Methane's lifetime in the atmosphere is much shorter than that of carbon dioxide, but it traps more heat from the earth's surface while there, stopping it from going into space. The warming impact of methane is 28 to 34 times greater than that of carbon dioxide over the first 100 years after it reaches the atmosphere, but 84 to 86 times over the first 20 years.

Livestock produce methane – and that means not just cows, but also sheep, goats, camels, etc. – as do our farming activities. We also produce methane because we bury waste in landfills, which leak the gas as it decomposes. Most importantly, however, we produce methane because the oil and gas industry leaks it from absolutely everywhere and then generates more by burning natural gas.

Millions, maybe hundreds of millions of years ago, and over long periods of time, the remains of plants and animals built up in thick layers on the earth's surface and beneath the oceans and mixed with sand, silt and calcium carbonate. Over time, these layers were buried under more and more sand, silt and rock. When pressure and heat is pumped into this carbon- and hydrogen-rich material, it forms coal, oil and natural gas.

We've been using natural gas since ancient times. The Oracle of Delphi, on Mount Parnassus in Greece, owed its mystical reputation a thousand years before Christ to natural gas that seeped through the rocks, affecting the minds of Pythia and her devotees when they inhaled it. Five hundred years after that, the Chinese were transporting natural gas through bamboo pipelines before burning it to desalinate seawater and make it drinkable. By AD 100, the Persians were using natural gas in their homes.

The first commercial use of natural gas occurred in England, where in 1785 it was produced from coal and used to light houses and streets. Three decades later, in 1816, Maryland in the US followed suit, the first city in the United States to do so. In 1904, gas was used to provide central heating and large-scale supplies of hot water in London, changing the lives of millions for the better. Then we started using it to generate electricity. In 1940, the first natural gas turbine was born, in Neuchâtel, Switzerland. From then on, the use of gas for electricity grew exponentially.

Today, combined cycle power plants, which produce electricity by burning gas then capturing the waste heat from the gas turbine to generate extra power from a steam turbine, operate with greater efficiency and lower emissions than other types of fossil fuel plants. They've been so successful that natural gas plants today supply more than half the energy consumed in residences and commercial applications in the US, which they do while producing half the carbon dioxide, a third of the nitrogen oxide and 1 per cent of the sulphur oxide of the average coal-fired plant. In other words, natural gas still kills you, but more slowly.

We find natural gas in various forms in the rocks beneath us; it's sometimes called 'conventional gas' or 'shale gas' or 'associated natural gas', but it's basically all the same. The only difference is where you get it from. You drill an oil and gas well, or a gas well, or in the case of fracking, you inject an enormous amount of chemicals and sand into the rocks to release the gas. You then extract it and, through a process of separation, isolate it. After this, it goes into processing plants, where non-hydrocarbon gases are removed by venting them or flaring them, which is another way of saying dumping them into the atmosphere.

From the gas-processing plant, the fuel goes into a compressor station and eventually to a natural gas company, which either sells it to consumers through pipes or stores it in liquid form to ship it somewhere else.

The problem – as discussed earlier, in Chapter 9 – is that natural gas leaks methane at every single point in the cycle. That's the main reason why methane now comprises 15 per cent of all our greenhouse gas emissions and rising.

Independent monitoring of the world's largest publicly traded oil and gas companies (as well as national producers such as Saudi Aramco, Russia's Rosneft and PetroChina) reveals that their methane emissions are both routine and at least 60 per cent higher than they're disclosing. A 2019 *New York Times* investigation showed that 'immense' (their word, not mine) amounts of methane are escaping from oil and gas sites nationwide. They discovered this by photographing the secret pollution with a highly specialized camera in a tiny plane crammed with scientific equipment. (Oil and gas companies don't have to disclose their pollution and emissions, which is in and of itself scandalous.) These findings are getting more alarming with time. In a 2022 study, Stanford University researchers found that oil and gas operations in New Mexico's Permian Basin were leaking six times more methane than the Environmental Protection Agency believed to be the case.

Since the Industrial Revolution, human activity has caused the amount of methane in the atmosphere to double. Today, methane pollution is responsible for one quarter of the global warming we're living through.

But while there's nothing but bad news about oil and gas emissions, there is good news as far as livestock and farming are concerned. Here's why: if a farmer changes what their cows or sheep are fed, they can cut the methane they emit from their other ends. More and more farmers are doing exactly this. In California, for example, farmers are feeding cows a mixture of dried seaweed and molasses. In Spain, they're giving them onions. India has an entire national programme focused on optimizing cows' diets to reduce methane emissions, and doing so in a way that produces more milk. All over the world, farmers are also investing in methane digester systems, a simple engineering solution to capture the methane that builds up in manure tanks and then use it to produce electricity.

Importantly, we are getting much better at cultivating genuine meat by taking cells from animals (without causing them harm along the way) and nourishing them in a way that mimics what happens inside the animal's body. This is done in a bioreactor, where the cells are fed with basic nutrients such as vitamins, sugars and fats. There are today dozens of companies backed by billions of dollars producing cultivated meats in bioreactors. Their products are improving in taste, texture and appearance as hundreds of millions of additional dollars pour into the related scientific areas of cell biology and culture and tissue engineering.

In 2020, Singapore became the first country in the world to approve the sale of cultured meats when the Singapore Food Agency endorsed cultured chicken made by Eat Just, a San Francisco-based start-up that is backed by over $400 million of funding.

Shortly thereafter, cultured chicken featured on the menu of Club 1880 in Singapore, marking its world debut on a dinner table and realising Winston Churchill's 1931 prediction that one day 'we shall escape the absurdity of growing a whole chicken in order to eat the breast or wing, by growing these parts separately under a suitable medium'.

I tasted Eat Just's first Good Meat product (the brand under which the company produces cultivated chicken and beef) in 2022 at one of the limited dinners the company hosted while building its new Singapore production facility, which is scheduled to be completed in 2024. Three plant-based appetizers were served in an immersive setting: a private room with art and videos on three of the four walls highlighting the problems the company was tackling.

'Forest Floor', made with mushrooms, aubergine puree, Jerusalem artichokes, Brussels sprouts, parsnip crisps, edible flowers and pumpkin seeds, was a colourful imitation of what a prehistoric rooster might have foraged. This was followed by 'Fields of Corn', a dish made with charred corn bread, barbecued mushrooms, miso baby corn, creamed corn and puffed sorghum, a dig at the corn we produce to support industrial meat production. 'Flooded Future' was next: celeriac broth with pickled white fungus, seaweed and blue pea flowers, an ode to human-caused climate change. Finally the *pièce de résistance* arrived: chicken nuggets served American style (with maple waffles and hot sauce) and Chinese style (in a bao with hoisin sauce), in a wink to the two largest chicken producers in the world.

As hard as I tried, I couldn't distinguish these nuggets from the real thing. They were delicious. The future of meat was right there on my plate, waiting for consumers to accept it once regulatory approval came through. While the production of cultivated meat is energy-intensive, this is largely mitigated when the bioreactor is paired with renewable energy, delivering up to 92 per cent fewer carbon emissions than traditional meat, all without taking down a single forest or harming a single animal. No wonder there are already over 100 cultivated meat start-ups, which have collectively raised $2 billion between 2020 and 2022.

In addition, if we fight deforestation (as we are), we contribute to solving the beef problem. The top two commodities driving global deforestation are beef and soy, the latter ironically a key ingredient in an Impossible Burger (when you're not getting protein from a Wagyu or USDA Prime steak, you're highly likely to be getting it from a soya bean grown in a formerly forested part of the Amazon). While waiting for cultured meats to take off, however, fighting deforestation is about neither eating less beef nor consuming less soybean. It's about enforcing existing laws, curbing corruption and making the directors of big corporations that turn a blind eye to where their beef and soybean come from personally liable and prosecutable.

Proper governance of rainforests would increase the price of beef and soybeans, as less land would be made available for both and the market would then take care of ensuring we all consume less.

We should give kudos to the amazing Impossible Burger. It uses 87 per cent less water and 96 per cent less land, and results in almost 90 per cent fewer greenhouse gas emissions than a regular burger. But we need to keep our focus on the key driver of methane pollution: the oil and gas industry is the biggest contributor *by far*. It creates half or more of all methane pollution.

Humanity can certainly do better on sustainable agriculture and livestock – and it is trying hard, as discussed earlier in this chapter. We must also do much better fighting deforestation. The hydrocarbon sector, on the other hand, is doing worse because its entire purpose is to extract and sell oil and gas. It employs some of the world's brightest people, but trains them to think of methane (and environmental impacts in general) as being peripheral to their principal business. They're reminded constantly that they're at work to produce oil and gas and incentivized accordingly. Their employers want them to produce more of the stuff, more cheaply and faster, and pays them for these results. The environmental destruction left behind is a mere by-product, an afterthought to be managed by health and safety departments treated as cost centres.

We need systemic change, not behavioural change. Our legislators need to make exploration for new oil and gas illegal. No more new oil and gas (or coal, for that matter). None. This process has started but it needs to be accelerated significantly. A recently formed international alliance, the Beyond Oil and Gas Alliance, groups countries, states and provinces working together to phase out oil and gas production; its members already include countries such as Costa Rica, Denmark, France and Sweden, as well as sub-national jurisdictions Wales, California and Quebec.

In addition, we need the directors of big companies that turn a blind eye to whether their products and supply chains have contributed to deforestation, ecosystem losses and climate change to be personally liable for the environmental destruction they cause.

At the same time, we need to phase out our existing oil, gas and coal use as soon as possible, by having a *much* faster take-up of electric cars, electric bicycles, electric buses, electric scooters and, eventually, green-hydrogen-powered ships and planes. Plus lots and lots of good old-fashioned bicycles.

It's fine to stop eating beef if that's what you really want to do (it might even be healthier for you), but we all need to refocus on what is really causing man-made climate change: big bad oil, gas and coal companies and their enablers, and the big corporations responsible for deforestation.

Each of us can in fact make a small contribution to the environment just by eating healthy, local and seasonal produce where possible, while at the same time being careful about buying blindly into the organic food craze. 'Organic' means that no synthetic fertilizers, pesticides, herbicides or fungicides were used to grow the food. This usually means that organic food needs more land and more water to produce the same yield. That's bad. But it also means that we're being kinder to nature and avoiding the pollution caused by fertilizers and pesticides. That's good.

On balance, eating organic requires far greater awareness of where the food comes from, in order to minimize the risk of taking land and water resources away from those who need them, especially in poorer countries. That organic quinoa or avocado flown from Mexico to be sold in supermarkets around the world may well be crowding out land and water needed to feed local people in Mexico.

I'm all for organic food produced locally. But organic food flown long distances is a terrible waste of resources. Imports not only create more emissions due to transportation, but they're also, obviously, less fresh. Seasonal foods, if you can obtain them, are usually fresher and tastier. If schools started teaching this simple principle to our kids, I'm confident they would remind us constantly of it.

I'm neither an opponent nor a proponent of veganism or vegetarianism. But nor do I want to judge people based on what they eat. We're simply not going to turn nearly eight billion people into vegans or vegetarians, not now and not ever. Everybody should eat what they want. However, for their own well-being, they should eat healthily, and if they do, it will most likely be a lower-carbon, climate-change-fighting diet.

The focus on beef is a distraction within the debate. The carbon footprint of the food we waste ($408 billion worth in the US alone, grown on 18 per cent of US farmland with four trillion tons of water) exceeds that of the airline industry. Yet reducing food waste costs almost nothing, and cutting it in half over 30 years is the same as taking 2,570 coal-fired power plants offline – while helping to fight deforestation at the same time. We are producing an enormous amount of food that no one is eating, and reducing our food waste footprint is certainly the most effective climate action any individual can take in the food category.

So eat healthily, and cut back your food waste as much as you can. But if you have eco-anxiety, don't lose focus on the fight that really matters: phasing out our existing oil, gas and coal use as soon as possible and stopping deforestation. Both are driven by big corporations with no moral compass that desperately need to be more regulated to be responsible.

See also:
Chapter 9: Stinky Gas

20

Drive Electric Shamelessly – the Green Energy Revolution Is Here

An electrification revolution is unfolding, encompassing cars, buses, scooters and trucks. It is coupled with a twenty-first-century approach to mobility that prominently features redesigning cities away from cars. The result? Close to 100 per cent of all new cars, buses, scooters and trucks should be electric by 2030.

The signs are everywhere. The number of global public charging points has gone up from less than 200,000 in 2015 to about 22,300,000 in 2022, with no signs that this growth is slowing down any time soon.

In China, the largest automobile market in the world, plug-in cars are exploding in popularity and hit a high of 26 per cent market share in March 2022 (electric-only cars were at a 21 per cent share), up from a market share of 11 per cent a year earlier. At this pace, all new cars in China could be electric by 2025.

One of the biggest problems with driving electric as recently as 2019 was that you couldn't find a vehicle other than mainly Tesla and you had to wait in line. Today, however, a tsunami of electric models are arriving, made by almost all car manufacturers. These companies understood that either they were going to have to give everyone a choice to buy electric, or they were going to be bankrupt.

Then they went a big step further. In a period of just months, announcements came thick and fast, with Volvo, Audi and Ford Europe announcing they would stop producing all diesel and petrol-powered cars, trucks and SUVs by 2030. The last Mini with an internal combustion engine will arrive in 2025 and the company will go fully electric by 2030. Jaguar is first out the gate, going all electric from 2025. General Motors joins these companies in 2035. Investing in electric infrastructure has ramped up fast, showing that the green transformation is in full swing: global car manufacturers are investing $515 billion in electric vehicles and batteries, up 70 per cent compared to three years ago, with Volkswagen accounting for 21 per cent of the total.

Countries are also banning the sale of new petrol and diesel cars and vans with great urgency. Norway and South Korea say no more by 2025, Denmark, Iceland, Israel, the Netherlands, Slovenia, Sweden and Britain by 2030 and Canada, France, Portugal, Singapore, Spain and Sri Lanka by 2040. Following the Ukraine war, Germany joined the pack in 2022 (ditching its long-standing position of lobbying for exemptions for its car industry from proposed European Union legislation), agreeing to back an EU-wide mandate that only zero-emission vehicles can be sold from 2035 or even earlier.

Today, the electric revolution is happening abruptly and radically, just as it did when we moved from horses to cars, landlines to cell phones, and Blockbuster videos to Netflix. Had we not lost 40 years of technology development and R&D in the electrification sector because of the obfuscation and lobbying campaigns of Big Oil, with the active collusion of car companies, this fast transition to electric transportation would have been a lot less painful on those who work in the industry. Car manufacturers could have spent the last decade preparing for it instead of fighting it. But they get it now. Several companies, including BMW, Volkswagen, General Motors and Peugeot, are also manufacturing electric scooters and electric bicycles, having seen where urban transport was headed, while car and scooter sharing companies are exploding in popularity.

Electric transport has another hidden benefit. The batteries in cars, trucks and buses can enable a full clean-energy power grid because they can support the grid rather than just taking power from it. Vehicle-to-grid technology, or V2G, allows a plug-in vehicle or electric vehicle to act as a form of energy storage, letting electricity flow from the car to the distribution network and back. As an example of how this might work, a family could leave their home completely off the grid and entirely powered by their Tesla Cybertruck while they are off on a two-week holiday. Once vehicle-to-grid technology scales up (it's on its way, with start-ups in that sector raising $400 million in 2021, up from $232 million in 2020), we will be able to shut down 45 natural gas power plants for every three million electric cars on the road. Add to that the fact that the electricity network worldwide is greening rapidly because of how fast we are adding renewable energy, and you can see that electrification is basically unstoppable today.

Forecasts that just 30–50 per cent of new car sales will be electric by 2030 miss the mark, because they suffer from cognitive rigidity and assume the consumer is an idiot. Why would anyone buy a non-electric car when they know its resale value will plummet dramatically, in many cases to zero, because of bans on new petrol cars combined with rising awareness of their environmental destruction? The writing has been on the wall for some time. The stock price of TotalEnergies, the French oil major, peaked in 2007. The stock price of Shell peaked in 2008. The stock price of BP peaked in 2010. These three companies haven't seen the same level of stock market capitalization since, and most likely will never see it again. ExxonMobil peaked in 2014, because the US markets took longer to digest the scale of the coming disruption.

The stakes are enormous. Out of approximately $5 trillion in petroleum sales per year globally, at least 25 per cent is dependent on what type of car we drive. The gasoline that we use in our cars is the most consumed petroleum product in the United States, for example, where it accounts for 45 per cent of total oil consumption, while in the UK, more than half of oil is consumed for road transport. That's before considering diesel fuel, which is used in heavy construction equipment, trucks, buses, tractors, boats, trains, some cars and electricity generators. Norway is a case in point: electric vehicles make up 15 per cent of cars on the road in the country in 2022, and this has resulted in the country's oil demand dropping 10 per cent from 2011 levels.

Naturally, rearguard action by vested interests is fierce. It is principally via a large amount of noise directed against the renewable energy industry and electrification, in particular against solar, wind and battery power. Years ago, solar and wind energy were criticized because of 'intermittency'. For example, detractors of clean energy would point out that the sun doesn't shine all the time, or the wind doesn't blow all the time, therefore we have to have coal, we have to have gas, we have to have oil. It turned out we didn't, because we can store renewable energy cost-efficiently (with battery costs down 90 per cent in 10 years and continuing to decrease); it costs a lot less to produce; and it is much cleaner.

Then it was birds. Wind turbines were to be avoided at all costs because they killed enormous numbers of birds. Wind power facilities can of course harm birds through direct collisions, but the industry is at least aware of this and has been making efforts for years to site and operate turbines in such a way as to minimize and mitigate their impacts on wildlife. It also turned out that the number one killer of birds is cats, followed by collisions with buildings, but no one was complaining about buildings or cats. We then learned that wind energy actually helps birds on a global scale by curbing climate change.

Then there was the cost objection, which was propagated by not only oil, gas and coal companies and lobbying machines but also for over a decade by the International Energy Agency, the world's top energy body. But guess what? It turns out that solar and wind power costs kept dropping until they became cheaper than coal, gas and oil-fired power, without even taking into account the environmental destruction from the fossil fuels, which for some reason is conveniently forgotten. In 2020, the turnaround was so complete that the International Energy Agency even declared solar power the cheapest electricity in history.

We've now moved on to a more subtle objection, focused on waste and mining, which goes something like this: 'Oh my God, what are we going to do with all those solar panels and wind turbines and batteries at the end of their lives?' Or 'Oh my God, what about the mining practices to get the cobalt and lithium we need for electric vehicles and the clean energy transformation?'

On 12,000 hectares (120 square kilometres) of a cattle station in Australia, a massive project is beginning to take shape, the most ambitious renewable energy project anywhere around the world when it was announced. Its sponsors are planning to install 22 million solar panels in order to produce 10 gigawatts of solar energy (about what you need to power the state of Connecticut, or Ireland), and it's going to cost $20 billion or more. It's linked into a 3,700-kilometre undersea cable from Australia through Indonesia to Singapore and is going to start exporting electricity and perhaps hydrogen in 2027. One of its headline goals is to supply 20 per cent of Singapore's electricity. It's also an easy target for attacks on renewable energy: what do we do with these 22 million solar panels at the end of their life?

This has become the go-to objection to renewable energy and electrification in the age of the great oil and gas industry panic, as study after study (there have been well over 100) documents that 100 per cent clean electricity from a combination of solar, wind, water and batteries is not only physically possible and economically affordable, but also poised to power the deepest disruption of the energy sector in over 200 years.

This is also being enshrined in law in many countries and states. California, for example, the world's fifth largest economy, has enacted a requirement that the state must be powered by 50 per cent renewables by 2026, 60 per cent by 2030 and 100 per cent by 2045, which means zero coal, zero oil and zero gas in its energy grid. California enacted this law because the intermittency and variability of renewables can be managed and the whole can work together. Their waste can and will be managed sustainably too.

Solar panels last about 30 years. In the EU, the recycling of solar panels, as well as that of batteries and inverters (the equipment that standardizes the flow of electricity a solar panel generates in order for it to be used by the grid), is mandatory by law. This will inevitably soon be the case worldwide. The vacuum in US federal leadership when it comes to solar recycling is on its way to being filled. Meanwhile, individual states and the solar panel manufacturers themselves are on the case (the largest US solar panel manufacturer, for example, has developed in-house recycling capabilities). Extended producer responsibility – making the manufacturers of future waste responsible for disposal on behalf of their customers – is also expanding. China started its own solar panel recycling and safety disposal research programme and has already drafted various rules and regulations to manage the recycling loop, with more on the way each year. In Australia, the state of Victoria banned dumping all e-waste, including solar panels, while New South Wales is implementing systems to recycle solar panels at the end of their lives.

The common belief that solar panels cannot be recycled is a myth. On the contrary, they are well on their way to integrating into the circular economy, with every piece being used and reused. Because solar cell manufacturers are bound by law in the EU to fulfil specific requirements and recycling standards, technologies to recycle solar panels are already everywhere. Some of them even reach an astonishing 96 per cent recycling efficiency. When did you ever hear an oil and gas pipeline or refinery talk about recycling anything?

Solar recycling is in fact creating more jobs and billions of dollars in recoverable value – so much so that we will be able to produce two billion new solar panels without the need to invest in any new raw materials whatsoever. That's enough to generate 630 gigawatts of energy just from reusing materials, more than 45 times the installed solar power capacity of France.

Wind turbines are a similar story. Right now, 85 to 90 per cent of the total mass of wind turbines can be recycled, including the steel, the copper wire, the electronics and the gearing. There is, admittedly, a problem. The blades are challenging, because they are manufactured from complex composite materials in order to withstand hurricanes, for example, and to be lighter and more durable. This makes them difficult to dispose of. In addition, some of them are as long as a football field and are difficult to move around.

The DNA of the industry is once again riding to the rescue, by building a circular economy around its products, and the world's first 100 per cent recyclable wind turbine blades are here: in 2021, Siemens Gamesa, one of the world's largest manufacturers of wind turbines, produced the first offshore wind turbine blades that can be fully recycled, including for use in the automotive industry or in consumer goods. The following year, a consortium of French and American companies produced the first prototype of its own 100 per cent recyclable wind turbine blade. A Danish start-up has also found a way to crush these blades, turning what is basically an ultra-resistant mix of fibreglass and glue into barriers designed to block noise from highways and factories. Other start-ups are developing methods to break down the blades and press them into pallets and fibreboards, which are then used for flooring and walls. Finally, Vestas, a Danish manufacturer of wind turbines, announced that their blades will be 100 per cent recyclable by 2024 through a new technology that will disassemble the blade's composite material at the end of its life.

Regulation is also helping. In the European Union, which strictly regulates what can go into landfill, some blades are already burned in kilns that create cement, or in power plants. The industry and its proponents are clearly focused on finding solutions. But in any case, this issue of the fate of wind turbines after their useful life is over could benefit from some perspective. Between 85 and 90 per cent of the total mass of wind turbines (and soon 100 per cent) is already recycled; the only residual problem is the blades, but these are already landfill-safe, unlike the waste from oil, gas and coal, for example. In addition, all the wind turbine blades that will be discarded worldwide through to 2050 are only equal to 0.01 per cent of the municipal solid waste we are already sending to landfills today.

Bear in mind that these are new problems that are already being solved, whereas what renewable energy is replacing – i.e. coal, oil and gas – is leaving us with a waste tsunami of plastic, pipelines and abandoned rigs leaking methane, all of which pollute our air, water, lungs and food chain, and none of which the fossil fuel industry is paying for. We need to see the forest for the trees. The environmental disaster lurking within the energy transition is only about what the oil and gas industry is leaving behind. Comparatively, the clean energy industry is boringly responsible.

It's not just solar panels and wind turbines that are about to be full members of the circular economy. Batteries are, too. Over the course of its life, a Tesla Cybertruck's battery will last over 1.5 million km (equivalent to 200 years). Ninety-five per cent of its lithium-ion battery components can be turned, right now, into new batteries or used in other industries. The plastic, aluminium and copper are extracted, separated and sent to their own recycling processes. What's left, chemical and mineral components referred to as the 'black mass', is treated on an industrial scale. Simply applying heat to extract the cobalt, nickel and copper, for example, gets you to a 48 per cent recycling level. More sophisticated methods then increase that to 95 per cent.

One of the co-founders of Tesla, J. B. Straubel, set up Redwood Materials to become the world's top battery recycling company, because it's abundantly clear that there'll be a point where if we can recycle all the batteries out there, we won't need to mine much any more. The company recycles enough batteries to power tens of thousands of electric cars. They're funded to scale up and are just one important participant in creating a circular supply chain to recycle the scrap from battery cell production, as well as from phones and laptops, power tools and power banks, scooters and electric bicycles, all of which require these batteries.

Redwood Materials and companies of its ilk collect the scrap from consumer electronics and battery cell manufacturers, then process this to extract cobalt, nickel and lithium, which they sell on to customers, allowing the supply chain of battery manufacturers to benefit from recycled minerals rather than newly mined ones. That's the clean energy industry again taking the lead on creating a circular economy, in contrast to phone manufacturers, for example, who put very little effort into recycling their batteries but seem to escape scrutiny. Mobile phones have the same lithium-ion batteries and 150 million are discarded each year in the United States alone.

Clean tech is a young industry. Its waste problem is mostly in the future, given the 15-to-30-year lifespan of solar panels, wind turbines and batteries. The industry is already developing and implementing solutions. Its DNA has a circular economy built into it, and the facts speak for themselves. Even when it comes to mining.

Cobalt is used to make batteries for electric cars. Fifty per cent of the world's production of that mineral comes from the Democratic Republic of Congo, which has all sorts of problems, including armed conflict, illegal mining, human rights abuses, child labour and environmental issues. Similarly, the extraction of lithium has significant environmental and social impacts, including causing water and air pollution and depletion. Mining copper, ammonium, nickel and graphite, some of the other raw materials used in lithium-ion batteries, is not done very well either and suffers from some of the same challenges.

The clean tech industry is therefore working on solutions to allow it to make batteries without cobalt, while trying to improve the appalling practices of the mining industry. Tesla, for example, has announced an alternative to cobalt and nickel batteries. These lithium-iron-phosphate batteries are already widely used in electric cars and buses in China because they are cheaper than those with nickel and cobalt. We can mine responsibly, and the clean tech industry is increasingly policing itself to try to do just that. One area of focus is pushing for new laws to create a regulatory framework for lithium and all other mining.

The alternative in any case is dirty cars powering dirty air worldwide. Burning fossil fuels is about much more than polluting the atmosphere with greenhouse gas emissions because of the local pollutants released at the same time. These, a mix of gases and tiny particulate matter, lead to almost nine million people dying each year.

There is no question that tapping into natural resources and extracting materials from the earth or the oceans damages the environment. The redeeming characteristic of the solar, wind, electric car and battery industries is that they understand this impact on resources and seek to avoid it, unlike their predecessors, the dirty fossil fuel industries they are replacing. That's why the green economy pushes us to think bigger and to never stop.

21

Green Bonds Do More Harm Than Good

Investors often say the right thing in public, but their actions continue to favour the old world of polluting fossil fuels and lax sustainability. They cheer when a company cares about the environment – but only as long as the share price is up.

Danone is a French food company that's been around for 100 years. It started with one product, yoghurt, sold in pharmacies for its health benefits and built on cultures from the world-famous Pasteur Institute in Paris. Now 100,000 employees strong, it sells about $30 billion worth of yoghurt, other dairy products, plant-based products, baby and medical foods and water. Its brands are everywhere, from the ubiquitous Evian to the fashionable Alpro plant-based products made from soy, almonds, hazelnuts, cashews, rice, oats and coconut. Its corporate flag flies in 55 countries. It sports respectable profits, operating margins and return on equity, in line with those of its peers. In early 2021, it was probably the most exemplary global multinational from an environmental, governance and social perspective. Its legal status is that of an *enterprise à mission*, or purpose-driven company. This means that while it is for-profit, it takes into account the interests of more than just its shareholders when it does business – including other stakeholders such as employees and communities, as well as the environment.

Danone is committed to be powered 100 per cent by renewable energy by 2030 and was already at 50 per cent in 2020. It's a founding member of the One Planet for Biodiversity business coalition on biodiversity, with a specific focus on agriculture and eliminating deforestation. It was the first global company to show its earnings after taking into account all the carbon emissions it is responsible for, trailblazing the concept of carbon-adjusted earnings per share. This approach is rare: at the moment, most corporate pollution is penalty-free. Companies don't pay for the carbon emissions they generate, unless they are subject to a carbon price, which very few are. They don't pay either for the recycling and waste management infrastructures societies have to build to deal with waste products. In a climate emergency, profits that are adjusted according to the cost of a company's carbon emissions are imperative to show their real earnings (i.e. after their environmental impact is taken into account), but no one disclosed them until Danone did. And no one has followed suit.

Danone is also trying to lead the way in regenerative agriculture. It is pushing for soil health programmes to improve how much carbon we keep in the ground while increasing yields. It's invested millions of dollars in research efforts to reduce the use of harmful chemicals in agriculture, avoid deforestation, protect and restore biodiversity and benefit farms and communities along the way by allowing them to eschew short-term goals and focus on their staying power. The alternative, industrialized agriculture, doesn't do any of these things and instead relies on massive quantities of chemical inputs like fertilizers, pesticides and antibiotics while paying little attention to deforestation and biodiversity.

But then it went wrong.

In March 2021, Emmanuel Faber, the CEO who made Danone a poster child of sustainable capitalism, was fired in a boardroom coup engineered by activist shareholders. Danone's stock had underperformed compared to that of its peers, Nestlé and Unilever, and the investors weren't happy. During a year stamped with the seal of a global pandemic, Nestlé's stock outperformed Danone's by 14 per cent, while Unilever's outperformed it by 4 per cent. Over five years, Nestlé's stock outperformed Danone's by 60 per cent, Unilever's by 40 per cent. On these metrics, Danone had lagged.

Naturally it isn't easy to determine who are the good corporate actors and who are the bad ones. We don't know the exact biodiversity footprint of these three companies and how they compare. How much deforestation is each responsible for? What is the burden imposed on society by all the plastic bottles they create, and are they paying for any of it? It's easier to look at the share price.

Capital markets don't currently give chief executives the time to build a purpose-driven business with a small environmental footprint. CEOs tend to be judged on short-term earnings devoid of environmental metrics and are hired and fired on that basis. Worse, they are penalized when they break norms; for example, when they turn their businesses into purpose-driven companies that aren't necessarily prioritizing paying dividends to shareholders ahead of taking care of their employees, communities and the environment.

Unilever itself had gone through the very same experience a few years earlier. The ubiquitous consumer goods company is the world's largest producer of soap, and manufactures much of what's in our cupboards (including energy and soft drinks, baby food, ice cream, tea and coffee, cleaning agents, pet food, bottled water, chewing gum, frozen pizza, juice and personal care products). Paul Polman, Unilever's CEO from 2009 to 2019, positioned himself as a corporate green leader by trying to bake environmental awareness into the company's DNA. It was a tough task: Coca-Cola creates the biggest plastic pollution footprint in the world, followed by PepsiCo, then Nestlé. Unilever continues (to this day) to be a strong number four, with a plastic pollution footprint of 70,000 tons per year – that's enough to fill 11 football pitches a day. And that's in just six countries: China, India, the Philippines, Brazil, Mexico and Nigeria.

For part of Polman's tenure, the capital markets retaliated by discounting Unilever's shares relative to its peers, in effect punishing the company for its growing green credentials. This proved the perfect window for a takeover offer in 2017, led by the chairman of Kraft Heinz, a partner at 3G Capital – according to the *Financial Times*, a group known for cutting jobs. Institutional investors, however, cheered on 3G's austere approach to management discipline and its indifference to sustainability. (Kraft Heinz didn't bother publishing a sustainability report, for example, until 2019, and its shareholders didn't appear to care.) Kraft Heinz's attempts at taking over Unilever were ultimately spurned, but a similar outfit could be successful in the future if institutional investors continue to make Unilever pay for trying to do good.

NRG Energy, the American utility giant, also attempted a transformation; this time from fossil fuel heavyweight to renewable energy powerhouse, under the leadership of another visionary CEO, David Crane. As far back as 2010, it started investing billions of dollars generated from fossil fuel plants in wind, solar, electric charging stations and energy storage. But institutional investors didn't buy into the new strategy: investors specializing in the electricity sector didn't understand it and didn't back it. NRG's stock crashed, the CEO was shown the door in 2015, and aggressive investors, looking for a short-term return, tried to break up the company.

In Europe, another far-sighted CEO, Gérard Mestrallet of the French utility giant Engie, put in motion a variant of the plan NRG had enacted. Engie embarked on a massive transformation from fossil fuel monster to green energy angel, during which it sold its large coal plants in Indonesia and India as well as its gas-fuelled US power generation portfolio. Simultaneously, it attempted to execute a global plan to replace its polluting fleet of power plants with clean energy assets, decarbonizing as fast as a large organization can.

The result? Its stock price went into progressive atrophy, and two or three CEOs later, the French government had to step in to temper the company's renewed enthusiasm for shale gas (as opposed to the green transformation it had previously embarked on) by stopping it from importing $7 billion of liquefied natural gas from the US and reminding it that it must seek cleaner supplies.

The experience of CEOs at Danone, Unilever, NRG and

Engie is that doing better by doing good doesn't go down well with international investors. Instead, we are being distracted by new financial instruments that seem to generate a fair amount of noise but are ineffective and may be doing more harm than good. Take green bonds and sustainable bonds. They're effectively IOUs issued by companies, banks, cities or countries that promise that if you give them the money they want to raise via the bond, they'll use it to make climate-friendly or environmentally friendly investments. On the face of it, you would think that's a good thing. They are intended to ensure the projects they fund have a positive environmental impact. In reality, they're being used in far too many cases to distract us from the fact that companies are being punished for doing good, and as an excuse to do far too little, while masking environmental destruction.

Royal Bank of Canada doesn't even pretend that its green bonds are going to be used to fund green projects. It has said very clearly that they may be used to finance oil and gas companies. In addition, when banks issue green bonds, the money raised simply goes into their balance sheet. Because all money is the same, in reality nobody's got a clue where it's going: the bank can tell us anything they want and we can't disprove it.

Globally, green bonds have reached their most substantial milestone yet, with over $1 trillion in cumulative issuance since the inception of the market in 2007. That sounds like a big number, but it isn't: green bonds account for a measly 1 per cent (if that) of all issued bonds, a market sized at approximately $130 trillion. In other words, over 99 per cent of the bond market is functioning very nicely with no regard whatsoever for whether the money raised is exacerbating climate change, destroying the environment, polluting rivers and waterways or endangering public health; or indeed, all of the above.

It is arguable that green bonds do more harm than good: they power a feel-good bandwagon ridden by bankers, publicists, lawyers, consultants, companies and countries to make us believe they're doing something substantial to save the world, when in fact they're doing nothing of the sort. The sad truth is that companies, and countries, don't have to do anything green at all to issue a green bond.

In 2016 and again in 2018 and 2019, Poland, for example, raised billions in the form of green bonds to fund token projects that distracted from the fact that its government's publicly stated policy was to promote coal and do whatever it could to slow the EU's march to a cleaner world. The integrity of the green bond market would have been protected had first-time green bond issuers been required to issue only green bonds once they'd issued their first. It's not okay for countries like Poland to raise green bonds for token projects, then raise other bonds to build coal plants, promote deforestation and otherwise ignore environmental and social standards.

The banks involved are apparently fine with all that, as are the investors buying the bonds. However, this is the antithesis of green investing: using green labels to obfuscate, hide environmental irresponsibility and conceal destruction.

Indonesia was the first Asian country to sell bonds labelled green, raising billions in the process. The small print, however, commits it to absolutely nothing other than vague green promises. This from a country better known for its out-of-control palm oil production driving massive rainforest loss and health hazards;

and for an energy sector addicted to oil and coal, even though Indonesia is endowed with plentiful sun and wind.

Compare Indonesia to Finland, for example. Finland, population 5.5 million, is an Arctic nation with about one third of its territory existing above the Arctic Circle, where it's very, very cold. Indonesia, population 268 million (that's 52 times more people than Finland), is a huge country with about 17,000 islands splashed across the equator, where it's very, very warm. But guess what? Finland's installed solar energy capacity is bigger than Indonesia's. That's the same Indonesia using the green bond markets to look good while acting bad.

Banks underwriting these green bonds and investors buying them are responsible for many other outrageous outcomes. There are green bonds funding coal labelled clean when there is no such thing as clean coal. There are green bonds issued by fossil fuel utilities and by oil companies. There are green bonds issued by banks with the money going no one knows where. 'Green' is being promoted as a tiny exception to the general rule, and even that exception is being abused.

At the moment, all you have to do to issue green bonds is to sign up to cheap and cheerful standards that are neither properly regulated nor enforced by anybody with regulatory power. Once you issue the green bond, there are no penalties if you subsequently break your green promises. There are no robust ways of measuring or verifying what the money is being used for.

There isn't even a consistent definition of what green is, and that suits the market very well, because banks, lawyers and service providers make very high fees from selling these bonds before going off and winning industry awards that their publicity machines leverage to hide everything else they're doing. Green bonds can even cost more than regular bonds, paying a higher interest rate or trading less favourably than their non-green twins.

In reality, all bonds should be green and we should be developing solutions to ensure that is the case. In other words, green bonds should be phased out. This will then ensure that the fig leaf that investors are using to feel good about themselves without making any difference whatsoever is removed.

We need to focus on ensuring that corporations, especially large global ones, are forces for good. Today, even those trying hard, like Danone or Unilever, are restrained from truly making the world a better place because the money markets force them to focus on short-term earnings instead of on their environmental impact.

We will know that we are on the path to green capital and money markets when sector rankings are reshuffled according to how sustainable a business is, and executive pay is determined accordingly. In the consumer goods sector, using Nestlé, Unilever and Danone as examples, applying a sustainability lens would dictate that Danone – the highest rated from an environmental sustainability perspective – would have the cheapest cost of capital, while Nestlé would be punished in relative terms for its lax approach to making the world a better place. This would then allow Danone to trade at higher relative multiples, enabling it to put cheaper capital to work and grow faster. It's not happening yet, but we don't have time to wait.

22

Tinker, Lawyer, Banker, Fry

'The first thing we do, let's kill all the lawyers.' So says Dick the Butcher in William Shakespeare's *Henry VI Part 2*. It's one of Shakespeare's most famous lines. But what did he mean? There are many possible explanations, debated for hundreds of years. I like to think that he was criticizing how lawyers maintain the privileges of incumbency, wealth and power. Together with bankers and insurers, that's still what they are doing today, certainly in the field of climate change.

There is a pervasive assumption everywhere in the institutions that are custodians of our finances and personal affairs – banks, insurance companies and law firms – that we are supposed to trust them when it comes to climate change action. Recently, they have all been busy aligning their public pronouncements to the objectives of the 2015 Paris Agreement. That's because climate change has been accepted by global public opinion as a real and present danger. In the US, for example, public polls show that 65 per cent of voters agree that America should be doing more to address climate change, while 70 per cent support their country's participation in the Paris Agreement. These numbers are even higher in Europe, where 76 per cent of adults in the UK and 93 per cent of those in the EU see climate change as a serious problem.

The Paris Agreement is a legally binding international treaty on climate change that was adopted in 2015 and entered into force in 2016. Its goal was to limit global warming to 2° Celsius compared to pre-industrial levels. It was the first time all nations came together to combat climate change in an organized fashion, after decades of talking about it.

Panic had set in in the preceding years because emissions kept relentlessly increasing, and over time many countries were encountering the consequences of climate change. Agreement finally came when a fundamental compromise was made: every country would submit its national plan for climate action and would then be left to its own devices to implement it, or not. No enforcement mechanisms were made available to the wider group of nations and no penalties could be levied. The public shaming threat was about the most powerful enforcement tool available under the Paris Agreement.

Nonetheless, the commitment of various governments to climate action in many countries, coupled with pressure from public opinion, meant that the Paris Agreement took on a life of its own and became the compliance benchmark against which companies and financial institutions were measured. Aligning with it became the rule.

Alignment means two things. We have already locked in 1.5° Celsius warming above pre-industrialization time and will likely cross this threshold by 2030. Therefore the main consequence of being aligned with the Paris Agreement is zero expansion of fossil fuels extraction. The second consequence is that we must cut emissions by 50 per cent by 2030, then to zero by 2050, to even have a 50 per cent chance of limiting global warming to 2°. That's without relying on wishful thinking about carbon capture technology, which doesn't work, or other magic solutions.

Take geo-engineering, for example. This describes technologies that may stop or reverse global warming by diverting the rays of the sun or otherwise masking the effects of our greenhouse gas pollution. If it sounds like science fiction, it's because it is. Geo-engineering is talked up in order to have a good excuse to keep burning fossil fuels and increasing CO_2 concentration while peddling hope. As it happens, it requires lots of natural gas to get the CO_2 out of the air, thus perpetuating the use of fossil fuels. We also have no idea how geo-engineering will impact us and our ecosystems. It would be far more efficient not to pollute in the first place, of course. Simple, right?

Banks, insurance companies and law firms, sophisticated institutions all, are intimately familiar with the Paris Agreement. They know what it means, and their public pronouncements assure us that their businesses are aligned with its goals. Their actions, however, ignore it.

Let's start with the banks. Bank of America, for example, one of the largest banks in the world, says it takes an active public policy stance on local, national and global climate change issues, and states that its goal is to advance clean energy and a low carbon economy. J. P. Morgan has a page on its website called 'Our Paris Aligned Financing Commitment' and assures the public that its policies are in line with the Agreement. Citibank says it's going to spend an enormous amount of money (its customers', of course) on funding the green economy. ING proclaims that climate change is one of the biggest challenges of our time and that they are committed to finance the energy transition. Barclays says their ambition is to be a 'net zero emissions' bank by 2050. In and of itself, this is meaningless – there are no standard definitions of what 'net zero emissions' means, and no one enforcing it in any case. Theoretically, however, it should mean that Barclays' activities, including those it finances, will result in no net impact on the climate from greenhouse gas emissions by 2050. And so on.

Now let's look at what the banks are actually doing. Sixty global private-sector banks have financed fossil fuels with at least $3.8 trillion since the Paris Agreement in 2015, backing 2,100 fossil fuel companies, in direct contradiction with their public pronouncements. Banks from everywhere are involved, including the US, Europe, Japan, China and Canada. US banks in particular have excelled in the hypocrisy department. The biggest fossil fuels banker between 2015 and 2020, by far, was J. P. Morgan Chase, followed by its US peers Citibank, Wells Fargo and Bank of America. During that period, J. P. Morgan Chase took double-speak to another level. It was the number one banker of fossil fuel interests by a large margin every year, backing oil, gas and coal with $316 billion in total. It was all done in quite a sophisticated manner too: they artfully drafted Arctic protection policies, for example – a type of policy that banks introduce to rule out backing oil and gas exploration in sensitive ecosystems – yet they still funded companies actively expanding drilling for new fossil fuels in the Arctic. They also had coal policies to stop funding coal expansion, yet they still financed coal plants.

In fact, just one out of 150 top global banks, insurers and investors have oil and gas policies consistent with the Paris Agreement, according to the 2022 Oil and Gas Policy Tracker, a tool developed by Reclaim Finance, an NGO, to analyse fossil fuel policies adopted by financial actors.

The banks have successfully created a perception with the public that they are on the right side of history. In actual fact, their greed knows no bounds. Take tar sands oil. The tar sands is the name of a geographical area twice as big as France located in Canada's Alberta province, where the largest deposit in the world of oil sands – a mixture of sand, water, clay and a type of oil called bitumen – can be found. The sands are smack bang in the middle of Canada's boreal forest, home to hundreds of indigenous Canadian communities, a vast array of species of animals and plants, tree cover capturing carbon pollution and clean water filtered by massive wetlands. Tar sands oil is famously dirty because it's of a very heavy type that's difficult to reach and extract. As a result, it's the most polluting of any type of oil, generating climate-warming emissions per barrel of up to 60 per cent more than the lightest form of oil. In addition, the resource-intensive extraction and transportation of the oil destroys the local ecosystems and their communities. Yet banks have splurged more than $110 billion of financing on tar sands oil in the years following the Paris Agreement.

It's the same story with oil and gas in the Arctic, a territory claimed by eight countries – the US, Canada, Russia, Denmark (because Greenland, the autonomous Danish territory, is two thirds within the Arctic Circle), Norway, Sweden, Finland and Iceland. We've been plundering the Arctic since the late nineteenth century. It all began with the legendary Klondike gold rush from 1896 to 1899, which drove 100,000 prospectors to dig for gold in Yukon, a Canadian province. This was soon followed by coal mining. Exploring for oil and gas followed at scale in the 1960s, with large-scale extraction since the 1970s in the Russian, Canadian and Norwegian Arctic regions. The plundering is facilitated by banks, which provided a whopping $314 billion of funding to Arctic oil and gas expansion between 2016 and 2020.

Multiple banks continue to back other climate-destroying activities, including offshore oil and gas (extracting oil from the oceans), fracking for natural gas, and natural gas in general. While some banks have stopped funding Arctic and tar sands oil, that's not good enough. Even though it's accepted today that any new oil and gas expansion is only going to make the climate crisis worse, only one bank, France's La Banque Postale, has stopped funding all forms of fossil fuel extraction. Most banks aren't entirely withdrawing support for coal mining and coal-fired power plants either: even when not providing the coal industry with finance directly, they continue to support it by advising coal companies on how to get funding from others, and earning fat fees in the process.

Fossil fuels and banks are joined at the hip. The impact of this marriage is extremely concerning. In 2019, the governor of the Bank of England admitted that financial institutions were funding enough carbon-intensive projects and companies to guarantee a 4° rise in global warming above pre-industrial levels. It's clear that we can't trust the banks on climate: their financial incentives are designed to ignore any long-term threats because they get paid based on their performance over a twelve-month period.

Instead, regulators must force them to measure and disclose climate risk in their assets (including those of the corporate and sovereign clients to which they are exposed), then price this risk. They must then be obliged to set clear, transparent short-term targets to phase out this climate impact. This has become obvious to banking regulators. The European Central Bank (ECB), barely one week after the 2021 Glasgow climate talks ended, released its first assessment of European banks' preparedness to deal with climate and environmental risks, covering 112 banks with assets of €24 trillion. Not a single bank was even close to meeting its expectations and over half had no concrete actions at all that the ECB could find in terms of embedding climate risks in their business strategy.

The Bank for International Settlements, the central banks' central bank, similarly condemned greenwashing in the industry, finding for example that banks labelling themselves as green are nothing of the sort. However, as the scrutiny of net zero commitments increases, calls for transparency do as well. National regulators now need to force transparency by turning the recommendations of the Task Force for Climate-Related Financial Disclosures (TCFD), an international scheme created in 2015, into obligatory disclosure requirements for the banking industry. Regulators know what needs to be done (the TCFD recommendations, for example, are comprehensive and extensive), but they aren't doing it robustly enough. Legislators therefore need to step in. The UK is likely to become the first of the world's 20 largest economies to promulgate legislation requiring the disclosure of climate-related risks after it introduced an amendment to its Companies Act.

Climate change is not just an existential risk, it's a systemic financial risk as well: every bank in the world faces the potential for dramatic losses from the failure of their fossil fuel clients. The banks also face losses from the physical and human impacts of climate change. Their broader loan portfolios are at risk too, because many, if not most, sectors use fossil fuels and therefore have different exposures to phasing them out. Consider airlines, for example. They spend 25 to 50 per cent of their money on fossil fuels. What happens when they stop buying oil and gas? Banks aren't modelling the impact. Or consider the building industry. What happens to oil and gas demand when the world switches to electric boilers and heat pumps? Are banks even modelling the impact of the electrification of transport on their loan books? They aren't. Fundamental economic changes are occurring beyond the investment horizon of their 12-month compensation packages and therefore these changes are broadly being ignored. Instead, banks continue to take our money and gamble it in sectors they know will inexorably shrink in the next few years. They will continue to do so as long as they get upfront fees, irrespective of how the investments and companies they back perform in the future.

Similarly, insurance companies are doing very little to fight climate change. While banks make a lot more noise, insurance companies shouldn't be forgotten. They are everywhere, embedded in the world economy. They are the companies that offer risk management to the rest of us, and they come in many stripes. Life insurance companies pay a death benefit as a lump sum when their beneficiaries die. Property and casualty companies insure against accidents of non-physical harm. Accident and health companies are the people who sell us our health insurance, for example.

There are other insurance companies that offer very specific policies, such as kidnapping and ransom, medical malpractice or professional liability insurance. Then there are the reinsurers, bigger insurance companies that insure insurance companies.

Many of us pay insurance companies automatically each month and don't give them a second thought. Then when we ask for a payout, we discover how difficult it is to get any money out of them. As a result, the insurance sector has accumulated over $30 trillion in assets under management, an enormous amount of financial power at approximately 10 per cent of all the money in the world.

Much of it is generated without any regard for the Paris Agreement. No new oil, gas or coal can be produced, processed or transported without insurance policies. Insurance companies, however, are delighted to write these. In 2020, Lloyd's of London, the world's biggest insurance market, made headlines when it announced that it was going to end new investments in coal, oil sands and Arctic energy by 2022; phase out existing investment in companies that derive 30 per cent or more of their revenues from coal, oil sands and Arctic energy by the end of 2025; and ask its members to phase out the renewal of existing cover to coal, oil sands and Arctic oil and gas by 2029. That announcement is a very good example of talking the talk but not walking the walk. Notice first that they've completely ignored the Paris Agreement, because they will continue to insure new oil and new gas. Even existing policies for coal, Arctic energy and oil sands are just fine until 2030, by which time atmospheric CO_2 concentration is going to be at 450 parts per million and 2° Celsius of warming will be pretty much locked in.

Even reports analysing the insurance industry's performance on climate change are few and far between. A landmark one called 'Insuring our Future', published by a broad collection of NGOs, produced a scorecard on insurance and climate change and catalogued the industry's failure to do anything of note at all. The NGOs wrote to 30 leading insurers asking them questions about their policies on coal, oil and gas. They were completely ignored by 12 of the 30. It turns out that just like the banks, even insurers that are supposedly progressive on climate change – a minority – are still paying attention to coal and the dirtiest types of oil and gas. Only 3 insurers implemented policies to stop insuring some or all new oil and gas production, while 14 have limited insurance cover for tar sands. Most, however, continue to insure projects which expand oil and gas production, despite the clarion call from the International Energy Agency in 2021 that we can't have any new oil and gas or coal if the world is to meet the Paris climate target. Only one company, Australian insurer Suncorp, has ended cover for all new oil and gas production and can therefore say it's aligned with the Paris Agreement. All the others are missing in action.

Instead, they should all have completely stopped insuring any new coal projects and new expansion of coal, oil and gas, and announced they will phase out insurance for oil and gas companies after 2040 when no fossil fuels should be insurable.

There are some encouraging signs. Insurers are making more noise about discontinuing coal coverage, while the insurance available to the oil and gas industry is concentrated in just a few places: 10 companies alone cover 70 per cent of the market, which means that if two or three of them stopped insuring new oil and gas projects, insurance premiums would shoot up.

Citizen action is also bringing change, albeit against heavy corporate resistance. A proposed coal mine in Australia, initially known as Adani Carmichael then subsequently rebranded as Bravus Mining and Resources, is a carbon bomb that would produce 4.6 billion tons of CO_2 over its lifetime, roughly eight times Canada's annual CO_2 emissions. In addition, the mine's owner, Indian industrial giant Adani, wants to export its coal to burn it in a power plant in India, which would sell the power on to Bangladesh. This would ensure that three countries that together contain 20 per cent of the world's population remain addicted to coal for the foreseeable future. Adani Carmichael has been fought by activist groups in Australia for almost a decade, in a relentless environmental campaign that has resulted in over 40 insurance companies ruling out backing it. If this proposed coal project is not insured, it won't go ahead, and that's the best way to fight it. It has certainly been more effective than counting on Australian government action.

Another project wounded by citizen action is the Canadian government's Trans Mountain Pipeline, a pipeline intended to ensure that Canada can pump more tar sands oil and ship it to the US. Hundreds of thousands of people have signed a petition calling on insurers to end their support for this project, and many insurance companies have felt the heat and dropped out. This type of organized and focused action by individuals can be incredibly valuable to effect meaningful change.

Insurance is an industry that is supposed to keep us safe, prides itself on using actuarial science and scientific rigour, and acts as the anointed king of risk management. However, it has missed pricing climate risk into its policies – quite an oversight given that climate change is an existential threat to policy holders and to the industry.

Insurers should be actively revisiting their projections for oil and gas, especially the new oil and gas we can't have, and their premiums to these sectors should significantly increase in order to avoid major potential financial losses. If that happened, the cost of capital of these fossil fuel projects would increase; some would become uninsurable and many would either stop or be phased out. Meanwhile, the insurance industry is fuelling more climate change while investing our money in dangerous sectors. Because their assets under management represent such a huge number, $30 trillion, their exposure is of major relevance to our entire financial system.

Then there are the lawyers. Climate litigation cases are up tremendously in recent years. In 2017, 884 climate change cases were filed in 24 countries. By 2021, cases had nearly doubled to 1,550 cases filed in 38 countries. All touch on fundamental life-and-death issues. Some of the key trends underlying that litigation are the abuse of fundamental human rights; the non-enforcement of existing climate and environmental laws; efforts to stop fossil fuel extraction and use; the fact that corporations should be liable for climate harms but usually aren't; and advocacy for increased climate disclosures.

Stunningly, the big commercial law firms are rarely doing any of that legal work. It's mostly not-for-profit organizations and volunteers. The large law firms are busy maintaining the privileges of wealthy and powerful corporations, without regard for corporate actions that are harmful to earth, the environment, public health, clean water and every single one of us. According to the 2021 Law

Firm Climate Change Scorecard analysis published by the Law Students for Climate Accountability, such firms love to work on cases making climate change worse, and they do that by a ratio of ten to one compared to cases fighting climate change. Second, they like to help the fossil fuel industry, the petrochemicals industry, plastic polluters and the likes, also by a ratio of ten to one. Third, they are enthusiastic lobbyists for the destruction of the planet, anti-climate action, in the same ratio.

It's definitely time for law firms to reconsider how they represent one of the most destructive industries in history and what they're doing about climate inaction. Every law student, every lawyer should (please) take into account what their firms are up to as they plot their career. All they have to do is look under the hood of whichever law firm they're considering joining or at which they're already working. The 2021 Law Firm Climate Change Scorecard shows that out of the top 100 law firms assessed, only three received an 'A' climate score, while nine received a 'B'. Put another way, 88 per cent of these firms are actively contributing to the climate catastrophe and received a 'C' or 'D' or 'F', with 37 per cent receiving a flat 'F'. They really must improve. There is no case for the defence. There is a compelling one for the prosecution.

Every coal mine, every oil well, every gas pipeline, every petrochemicals plant is supported by a web of legal contracts. The law firms make all that activity possible, not only because they structure contracts in a way that minimizes liability for environmental harm such as leaking contaminated water, but also by helping orchestrate financing for these businesses, including the infrastructure, acquisitions and refinancing of fossil fuels and other harmful assets.

For example, Allen & Overy, Milbank, Norton Rose and Shearman & Sterling – all big, powerful global law firms – advised on the financing of a massive $4.3 billion coal plant in Indonesia, notwithstanding the fact that thousands of people in Jakarta were publicly protesting the project because of its local public health impacts and its contribution to worsening climate change.

Law firms also send lobbyists to Washington DC, to Brussels, to Tokyo, to advance and push for the fossil fuel industry's agenda with politicians and agencies. That's an expensive activity. Companies pay a lot of money for that lobbying and law firms take it on enthusiastically to help them retain trillions in subsidies for fossil fuels. Their efforts get liability shields into contracts so that the 'polluter pays' principle does not apply. In other words, their clients can pollute at will and leave it for future generations to clean up. They also seek to slow down or cancel legislation that would limit emissions.

The impact of this lobbying is far-reaching. For example, the US government says there are three million abandoned, unplugged oil and gas wells. These release methane and pollutants every day, and it has been left to local governments and communities to deal with the problem. But why did the oil and gas companies that built these wells get away without cleaning up their mess? A big reason is lawyers diluting liability laws, paid hundreds of millions of dollars in fees for their contribution.

The Dakota access pipeline – a mega 1,886-kilometre-long underground oil pipeline that began construction in 2016 and became operational a year later – is another case study of the role that law firms play in the development of a destructive fossil fuel infrastructure project we can't afford to have. All the lawyers involved knew that there had been massive protests against it by indigenous tribes, activists and locals because the pipeline threatened indigenous heritage, put tribes' water supply at risk and leaked methane gas. But at least eight top law firms in the US actively supported the developers of the pipeline every step of the way. They helped secure financing, lobbied for the oil companies behind the project and accessed the courts constantly to put protesters in jail and fight environmental protections, buying the fossil fuel industry time. Time we don't have.

Adding insult to injury, the lawyers then advertised their prowess. A partner at Norton Rose says in his biography on the law firm's website that he represented a consortium of energy companies behind the project, 'successfully defeating attempts in the US district court for the District of Columbia to enjoin the construction of the pipeline'. The law firms never lose. If the project goes ahead, they have happy clients. If the project doesn't go ahead, they get paid anyway.

Law firms choose their clients and don't have any excuses. They can also withdraw from representation based on any good cause. In addition, their own legal ethics require them to disclose when a client's actions may result in reasonably certain death or substantial bodily harm. Now surely they can put one and one together and understand that by promoting the expansion of fossil fuels, they are promoting actions that may result in death? Nor is the concept of the right to have legal representation relevant. Fossil fuel companies can always find a lawyer; they are endowed with big in-house legal teams. Law firms representing them are closing their eyes and collecting their fee – clean water, the environment, the climate and public health be damned. Their brands and reputations should suffer and other clients serious about acting on climate should take their business elsewhere.

Hypocrisy is rampant. Law firms advertise how responsible they are and tout their sustainability credentials in an endless stream of PR. For example, global law firm Slaughter and May announced that it has ambitious new targets to reduce the impact of climate change and that it's committed to set science-based targets aligned with the requirements of the Paris Agreement. But it's complete greenwashing: who cares about the carbon footprint of a law firm that doesn't manufacture anything? It's not material in any way compared to the impact its lawyers have working on just one transaction to expand oil and gas production. The law firm is destroying the planet – its clients include Shell, Maersk Oil and Premier Oil among many others – yet thinks carbon-offsetting the travel of its staff allows it to showcase its sustainability credentials.

Law firms should commit to rejecting new fossil fuel industry clients, new petrochemical clients and plastic polluters. They should also phase out current fossil fuel clients. They have an obligation to disagree with business models that condemn future generations to disaster and displacement. Clients can also help. Companies making net zero commitments; countries making net zero commitments; development finance agencies; multilateral agencies and others committed to fighting climate change should scrutinize their lawyers' credentials before hiring them.

Importantly, law firms rely on talent, and that talent also needs to ask more forcefully who their clients are. Employees of law firms, or for that matter banks and insurance companies, shouldn't turn into robots thinking only about billable hours and earned fees when it might be the devil they're sending an invoice to. They need to get their moral compass out and start using it again. There is an ongoing climate emergency, but apparently neither the banks, the insurers or the lawyers have noticed.

23

The ESG Con

Imagine. Climate change pummels Indonesia. Tsunamis. Floods. Landslides. Drought. Earthquakes. Sea levels rise. Indonesians start to leave. A few at first, then thousands and tens of thousands. Twenty million men, women and children, crowded onto makeshift boats, some more elaborate than others, pack the Singapore Strait in a line stretching to the Java Sea, all in need of shelter, food and water. Millions more head to Australia, overwhelmed by nature.

Similar events unfold in the Mediterranean at the gates of Europe, and on the Mexico–US border, where millions of Africans, Arabs, South East Asians and Latin American climate refugees turn up. What are Singapore, Australia, Europe, America going to do? Are they going to shoot them? Are they going to let them in? Are they going to finish Trump's wall, but this time continue it all around their countries?

Over the past 200 years, we've delivered amazing progress, in no small part because of how much oil, gas and coal we burned. We doubled global life expectancy. We almost conquered illiteracy.

We reduced infant mortality from 50 per cent in 1800 to 2.9 per cent globally today. Most people are walking around with more information in their pocket than Abraham Lincoln or Louis XIV had access to in their entire lives. But we cooked the books while building this progress: every single earnings figure for every single company in the world is wrong, because it isn't pricing in environmental destruction.

At the beginning of the Industrial Revolution, we did not know we were cooking the books. In 1882, the world's first coal-fired power station, the Edison Electric Light Station, was built and commissioned in London. It lit up an area of the city, powered the telegram service, and generally made people's lives better. We then built a global infrastructure based on burning oil, gas and coal, and improved many more lives and livelihoods at a beautiful moment in time.

However, 40 years ago, we did know. In fact, a United Nations body, the IPCC, says we've known about our impact on earth's climate for more than 100 years. But approximately 40 years ago, in 1982, oil company ExxonMobil's internal scientific papers showed beyond a doubt that CO_2 warmed the planet; and that when we extracted oil, gas and coal and then burned them, a specific amount of CO_2 was generated, which we could calculate precisely.

We did little about it, however, as evidenced by the fact that emissions are continuing to rise to this day. Instead, we put together ever more creative ways to cook the books. First, companies, governments and lobbyists buried the truth about our impact on the climate. This allowed them to avoid the liability for any environmental destruction stemming from the burning of oil, gas and coal. How could they be blamed if the problem did not exist?

This was documented by Inside Climate News in its 2015 series 'ExxonMobil, The Road Not Taken' (a finalist for the 2016 Pulitzer Prize for Public Service), after an eight-month investigation. This investigation used ExxonMobil's climate research, internal documents and files dating back to the 1970s and interviews with former employees to demonstrate its efforts to bury the evidence when it discovered that burning fossil fuels led to increasingly dramatic climate change. As discussed elsewhere, starting in 1989, ExxonMobil and its peers began to obfuscate, criticizing climate models as inherently unreliable (even though these were the models ExxonMobil's own scientists had used, without pushback at that time) and leading campaigns to weaken the emerging global consensus on global warming.

Then, from the mid 2000s, when the evidence became incontrovertible, a chorus of institutional investors, corporations and Big Oil adopted a pass-the-buck approach. They began assuring the public that they wanted nothing more than to fight climate change, but constantly complained about the impossibility, for all of them, of estimating their impact and doing anything about it unless governments taxed carbon or forced them to adopt a carbon price (which governments weren't doing effectively).

Institutional investors and fund managers used this strategy most effectively. In 2019, 86 per cent of UK fund managers, for example, called on oil companies to align their businesses with the climate goals of the Paris Agreement. In and of itself they were right to do so. However, there are two problems. The first is that half of these very same fund managers have zero policies in place that will align their own investments with the Paris climate goals, and just a fifth have sufficient policies in place. If you're a fund manager calling on oil companies to do what you're not doing, what's the point? But more importantly, they are not taking into account embedded climate risks when pricing stocks and bonds. They are just passing the buck. However, this is nothing in comparison to a new tool to cook the books: ESG.

ESG refers to the environmental, social and governance risks embedded in businesses. The idea is to catalogue information about the ESG risks in a company's business because these might affect its performance. Another way to think about ESG is that it's a tool to assess how sustainable a business is. Is it affecting the environment in material ways that are not taken into account? Does it have social issues? Does it treat its workers well? Does it use child labour? Is its governance transparent and robust? Does it have checks and balances on its management team?

For example, if you wanted to look at the value of real estate anywhere in the world, you would rely on financial statements that show the profitability of that real estate. There is nothing in these statements that takes into account the investment needed to defend the property from future sea level rises; the deforestation, if any, associated with building it; how the project's staff are treated; or the costs in terms of CO_2 emissions that have gone into the construction of the building and its current operation. ESG gives you a beautiful report cataloguing these risks without either putting a number on them or deducting their costs from the profitability.

In order to make directors feel better about cooking the books, everything not on a company's balance sheet or earnings goes into a bucket, which is then forgotten about. That bucket is ESG.

ESG has always been with us in one form or another, though as a fringe activity. In the 1800s, for example, religious beliefs were sporadically used to avoid certain investments. In the 1900s, social issues like the anti-Vietnam war sentiment or the anti-apartheid movement led to investing filters. Throughout, impact investing and socially responsible investing (terms used interchangeably today with ESG investing) were categorized as niche and used by a minority of investors. They were also deemed to be unconnected to financial or investment fundamentals, even though that was clearly wrong.

It took the Paris Agreement for ESG to go mainstream. When 200 countries signed one agreement backed by a scientific consensus, public consciousness moved dramatically. Individuals around the world opened their eyes and absorbed the fact that climate change was now a reality everywhere around them. They wanted to make a difference and flocked to ESG products, and investment managers and banks gleefully jumped on the ESG bandwagon to take advantage.

ESG is such a useful feel-good factor that it has become a huge business worldwide. From zero market share of assets under management five years ago, ESG-labelled funds today account for one out of three professionally managed dollars. Soon, 100 per cent of professionally managed assets under management worldwide will be ESG-labelled, because it's very hard to justify why any investment isn't taking ESG factors into account.

There are, however, at least four big problems with ESG. The first is that there are no universal, clear rules. Instead, there are extremely elaborate spreadsheets from a multitude of ESG framework providers that companies can arbitrage between in order to derive the best ESG profile they can. There is some good news on this front. Leading ESG standards organizations – voluntary NGOs that have issued ESG standards – are trying to collaborate so that there is less confusion. The International Financial Reporting Standards Foundation (IFRS) is also finally, some 30 years late, developing ESG standards.

The IFRS Foundation is the Pope of the global accountancy profession and develops the globally accepted accounting standards that accountants apply when they review and sign off on financial statements. Because 120 countries use these IFRS standards as the basis for the preparation of financial statements, adding ESG standards will probably give them a global audience. That is a very good thing. The European Union has also introduced rules that police investment products, forcing asset managers to be transparent about what they are doing in the ESG space, in order for European investors and citizens to be able to make informed decisions about their money – because it's clear at the moment that they cannot. The UK is working on doing the same thing. Perhaps most importantly, the US Securities and Exchange Commission (SEC) is also on the case, laying out an ambitious agenda promising to make environmental, social and governance issues central to its mission. Given that the SEC supervises the largest securities markets in the world, this is very encouraging.

That's all desperately needed, because the second problem with ESG is that inconsistencies and abuse are currently rife. For example, deforestation is a major driver of climate change as well as a significant factor in biodiversity loss. One would therefore think it should be a core focus for investors who want their capital to fund positive environmental change. One would be wrong.

Carbon Tracker, the green think tank, found that 78 per cent of mutual fund providers and 64 per cent of exchange-traded fund providers offered ESG investments, but not a single one of these funds specifically excluded deforestation risk, or actively priced climate risk. That's in spite of the fact that the Bank of England, for example, warns repeatedly that the global financial system could be left in perilous shape because $20 trillion of assets will be wiped out by climate change if institutions and companies fail to prepare properly. Twenty trillion dollars. Poof!

The inconsistencies are across the board. Out of 723 equity funds with over $330 billion in total net assets, 71 per cent of the ESG funds and 55 per cent of the climate funds assessed in 2021 were negatively aligned with Paris commitments; in other words they were financing an outcome contrary to the Agreement's specific aims.

The third problem is that epic greenwashing is everywhere. Out of 253 funds that switched to an ESG focus in 2020 in the United States, 87 per cent simply rebranded by adding words like 'sustainable' or 'ESG' or 'green' or 'climate' to their names. They didn't change their stock or bond holdings at that point. Another example is courtesy of the largest investment manager in the world, BlackRock, which launched its wonderfully named BlackRock US Carbon Transition Readiness exchange-traded fund in 2021. This broke funding records on its first trading day, raising $1.25 billion. But the fund was a mockery of ESG, with holdings such as ExxonMobil, Chevron, Conoco and Marathon Oil, all world champions of stopping decarbonization.

But all this is minor in comparison to ESG's fourth problem and biggest failing.

ESG is all about disclosure and reporting information. That's companies informing us about their environmental, social and governance risks. Disclosure and reporting are not the same as doing something, however. When auditors sign off on financial statements, pretty much anywhere in the world, they finish with something along these lines: 'In our opinion, the financial statements referred to above present fairly in all material respects the financial position and the results of operations and cash flows of XYZ company.' These words, 'present fairly in all material respects', may be some of the most dangerous in the English language today. They give us the impression that the numbers we read – and on which we base our entire $300–$500 trillion financial system – are correct, when they're not.

Disclosure may be sufficient for the 'G' in ESG, i.e. governance, because this is about corporate actions and behaviours. Disclosure may also be sufficient for the 'S', because this is mostly about human capital and the social issues companies deal with. These can both be catalogued, and we definitely need to understand them to assess corporate risk. For example, when Deliveroo, the food delivery company, went public in 2021, many institutional investors stayed away (and as a result, the share price tanked after the initial public offering) because they objected to both Deliveroo's governance (a dual-class structure allowed its CEO to retain control of the business for three years) and its labour practices (particularly how it treated its delivery drivers). Disclosure worked.

The 'E' in ESG, however, must be priced: environmental destruction must be quantified and accounted for. Destroying the environment and polluting free of charge cannot just be a disclosure and reporting issue.

These are not small details. The global corporate sector causes $7.6 trillion of damage to the environment each year from just three buckets: namely the emission of greenhouse gases, the disposal of unrecyclable waste in landfills and the discharge of untreated water.

ExxonMobil tells us its 2019 emissions were the equivalent of 730 million tons of CO_2. The social cost of carbon – in other words, how much damage in dollars one ton of CO_2 emitted into the atmosphere causes – is estimated to be anywhere from $50 to $200. The Biden administration, for example, currently puts it at $51. Let's suppose it is $50, which, multiplied by 730 million, is $36.5 billion.

In a good year, ExxonMobil makes about $20 billion in profits, but that's not taking into account the $36.5 billion of climate damage it causes. It doesn't include that because it's not paying for it, but we are, through the catastrophic climate change that we have already banked and continue to accrue. If we were not cooking the books, we would be able to see that ExxonMobil is in reality losing an enormous amount of money each year.

That's just ExxonMobil. Over the past 40 years and across the entire oil, gas and coal industries, the source of all man-made climate-warming emissions, we're talking about trillions of dollars that could have made our planet safer much faster. Instead, we used those trillions to keep looking for more gas, more coal and more oil, to keep producing more plastic and polluting stuff, and to keep paying everybody doing it very handsomely. This applies to every single company worldwide.

Take BASF, for example, the largest chemical producer in the world. By applying the same methodology, we can see that BASF is mis-stating its earnings by 67 per cent. Its real earnings, therefore, after accounting for its climate impact, are one third of what they appear to be.

Toyota, one of the largest car companies in the world, is mis-stating its earnings by 100 per cent and in reality has never been profitable after we take into account its environmental impact. Imagine how many electric cars and buses it could have manufactured in all this lost time had it been paying for its emissions.

Imagine a world in which the oil and gas companies led by ExxonMobil hadn't lied, obfuscated and lobbied for the past 40 years. Information about the harmful impacts of fossil fuels would have spread to governments and citizens around the world. The price of oil, gas and coal would have risen because we would not have cooked the books. The price of plastic would have risen. These price changes would have led us to consume less oil, gas and plastic all the way back to the 1980s. Electrification would have accelerated. Renewable dominance would be here already. Research and development into powering planes and ships cleanly and producing steel and cement without fossil fuels would have already delivered greener versions of these industries. So many fewer people would have died from pollution and climate change.

We have to fix ESG before it's too late so that companies price environmental destruction properly. Far more money would flow to the right places far faster, which is exactly what we need in a climate emergency. An electric boiler or heat pump would suddenly look like a bargain compared to keeping gas-fired ones. Green hydrogen would already be cheaper than fossil hydrogen. Greener steel would already be here, and perhaps even electric aircraft and ships would be too. Plastic bottles would have disappeared. What Bill Gates calls the 'green premium' (the extra amount you have to pay for a zero-carbon product compared to a fossil fuel equivalent; for example, the premium you pay for a biodegradable straw compared to a plastic one) would melt away and disappear. As a matter of fact, there is no green premium unless we cook the books.

We need governments and regulators to place a price on pollution so that everybody pays for their greenhouse gas emissions: every single company in the world. We've been waiting for 40 years and will probably continue to wait, except for limited pockets such as Europe, New Zealand and California where a price is already placed on environmental destruction, albeit for only a few sectors of the economy. While waiting for governments and regulators to muster the courage to do the right thing, investors must do this work on their own.

Fighting climate change effectively and decisively is not just about banning items like plastic straws. If we show the environmental footprint of plastic straws in financial statements, as well as the expense of collecting them and recycling them, they would be exposed for what they are: poison, and loss-making poison at that. We wouldn't need to ban them: they would just simply stop being manufactured.

We need good companies to take the lead. They should account in dollars for their environmental impact and show this in their financial statements. They should also compensate their executives based on these adjusted earnings, in order to align incentives with what's going on in the real world.

We need accountants to wake up and start questioning fictional financial statements that don't show carbon emissions and environmental destruction. At the moment, accountants are simply forgetting the financial costs of climate change and environmental destruction and signing off on mis-stated earnings.

This, however, is not just about governments, regulators, investors, companies and accountants. Individuals have the power to turn the tide and put a stop to the ESG con as well. They could start new investment research firms, for example, that issue restated earnings for the 1,000 largest companies in the world, showing their hidden environmental costs; or shed light on pervasive transgressions in the bond markets funding coal, gas and oil development.

Some have started trying, for example Swede Ulf Erlandsson is a bond vigilante. Backed by philanthropic foundations, he set up the Anthropocene Fixed Income Institute to publish research exposing the role of fixed-income investors and bankers in accelerating climate change by financing polluters. He uses a combination of analytics, public shaming and dialogue to get the banks and the bond markets to change their ways. Another pioneering changemaker, Briton Mark Campanale, cemented the terms 'carbon bubble' and 'stranded assets' into the vocabulary of finance through Carbon Tracker, a financial think tank focused on pushing investors to back the energy transition more forcefully by highlighting the investment and reputational risks involved in funding climate-destroying, carbon-intensive fossil fuels.

Every action contributes to building more momentum behind increasing the cost of money for polluters and spreading the word until real earnings become the basis on which companies are judged, executives are paid and stocks are traded.

Others can effect change from inside existing research houses and push them to begin publishing real, honest numbers; or reform the law firm, the accounting firm, the investment firm or the insurance firm where they work. Anything at all helps. Stopping cooking the books is probably the single most powerful step to steering the world away from a climate catastrophe.

24

Don't Worry (At All) About Bitcoin's Energy Use

Bashing Bitcoin, the blockchain and crypto miners for their energy usage is nonsense. You know the headlines and the statements I am talking about: 'Private-equity firm revives zombie fossil-fuel power plant to mine Bitcoin', or 'The cryptocurrency uses more energy than entire countries such as Sweden and Malaysia, according to researchers', or indeed, 'We are concerned about rapidly increasing use of fossil fuels for Bitcoin mining and transactions, especially coal'. This is all, in one word, noise and should be ignored.

Let's start with the basics. What is Bitcoin anyway? Is it useful? Aren't we better off with standard fiat money issued by national governments and banks? There are a few facts that are key when thinking about Bitcoin. The first is that fiat currency, what we commonly refer to as money, is clearly not enough to do everything we need to do. If it were, we wouldn't have been using gold as a store of value for thousands of years – and there are no signs of gold going anywhere. On the contrary, its value

231

rockets whenever uncertainty rises in the world.

Second, think about fiat money printed by governments for a moment. Fiat money experiences regular defaults, confiscations and loss of value (to zero sometimes) because of inflation or other factors. This is because it is run by monopolistic governments who aren't always competent, or acting in their country's, or its people's, best interests. Think Venezuela or Lebanon, for example, where millions have had their life's work confiscated and cancelled because of the gross negligence and corruption of their leaders.

With that context in mind, Bitcoin can be thought of as digital gold, except that as a gold substitute or complement, it is better. No one can break into your home or business and run away with a suitcase full of Bitcoin. In addition, you can send Bitcoin anywhere in the world in seconds, and it's infinitely divisible too. Yes, it can be very volatile in terms of its price fluctuations, but that's an evolutionary characteristic: we're still figuring it out. Meanwhile, an estimated 300 million people already hold crypto-currencies, 47 million of whom are in the United States alone. We're probably just a few years away from them numbering one billion.

Consider also that we have $10 trillion of gold around the world, versus about $1 trillion of Bitcoin. There's no reason why Bitcoin should go anywhere, just like gold. In fact, it may dwarf gold in market capitalization if it's more broadly adopted as a store of value. Bitcoin relies on something called a proof of work, a decentralized consensus mechanism that requires members of a network to solve cryptographic mathematical puzzles to prevent anybody from gaming the system. Unlike money, where a single entity, typically a central bank, has power over the system because of its monopoly on printing currency, a consensus mechanism is a complex governance system built to reach agreement between all participants, even though they don't know each other. Whoever solves the cryptographic puzzles first (the process of solving the puzzle is known as 'mining') is rewarded with bitcoins. All most of us need to know is that the mechanism is pretty much unbreakable and it's impossible to have fake bitcoins in the system.

The key characteristic of Bitcoin is that it consumes energy by design, in order to ensure the system is secure. In addition, this energy is distributed around the world across many countries, organizations, businesses and people – lots of machines mining bitcoins in multiple places – all of which act as a check on each other and which work together to make the system stronger.

With gold mining, you begin by destroying vast swathes of land to search for gold and develop a mine (while building physical infrastructure to access the mine and house and feed everybody involved), then dig underground to extract ore before having to process it, all while using twice the electricity you need to mine Bitcoin and sprinkling the process with human rights abuses, conflicts and environmental destruction, at great expense to local communities.

Bitcoin by contrast is mined from electricity, using maths, and is infinitely more environmentally friendly. Electricity is the mine and replaces human oversight to regulate financial activity. Miners receive coins that represent the electricity they've surrendered, and that is precisely its genius – storing electricity as an asset, for everyone's use. Counter-intuitively, perhaps, the more electricity the Bitcoin network consumes, the more robust Bitcoin is as an asset.

There are many additional things to love about Bitcoin. Today, the amount of energy we waste is phenomenal. Extracting and burning fossil fuels to produce electricity not only releases greenhouse gases and local air pollutants and consumes an enormous amount of water, but to top it all off, two thirds of the energy in the original fossil fuels is lost: fossil fuel plants burn three units of fuel to generate just one unit of electricity. Just capturing and using as fuel the natural gas burned or vented as a waste product of oil extraction would generate more than four times the electricity needed to power the entire Bitcoin network. There are rarely complaints about this gigantic waste, or widespread recognition that renewable wind and solar power burn no fuel, release no greenhouse gases or local pollutants and hardly use any water.

In addition to all this downstream waste, electricity is also wasted after it's been generated. That's called 'curtailment', a polite word to refer to the common practice of discarding electricity in the same way you might throw an aluminium can in a bin, except the can might be recycled whereas the discarded electricity is lost forever. The curtailment of electricity by utility companies is massive. China's alone is equivalent to all the electricity used by Bitcoin miners in one year.

We should think of crypto miners as the electric Pac-Man: they go where electricity is otherwise wasted and scoop it up, paying their way and doing electricity utilities a big favour. Electricity grids don't work at a constant rhythm. Often they have a lot of idle capacity when electricity use drops because of the time of day. That's capacity for which the producers aren't earning anything. Electric Pac-Man comes along and allows them to realize the full value of their assets, paying them for electricity during these times. Then, when electricity is needed elsewhere, it can be diverted away from Pac-Man. We've seen this in Texas, for example, when it experienced a power crisis in 2021. Electricity demand rose because of how cold it was, and the state's power plants suffered outages when their generating equipment and pipelines froze, fuel ran out and gas production was interrupted. Electricity to Bitcoin miners therefore became too expensive and they halted their mining.

In the case of countries uneconomically subsidizing dirty fossil fuels, such as Iran, Venezuela and Saudi Arabia, Bitcoin miners provide a societal service by consuming all their energy, ultimately punishing them. This played out in Kosovo in 2022, when newspaper headlines proclaimed 'panic' as Kosovo pulled the plug 'on its energy-guzzling Bitcoin miners'. The miners were there because Kosovo had the cheapest energy prices in Europe: 90 per cent of its domestic energy production comes from burning its rich reserves of lignite, one of the most polluting forms of coal, and the government subsidizes its fuel bills. Bitcoin miners therefore did exactly what they were there to do: they went to a country burning fossil fuels while subsidizing them and ate its energy up. When one of its two coal-fired power plants had an unforeseen shutdown, Kosovo had no spare electricity capacity (the Bitcoin miners had consumed it all) and scrambled to buy energy from the international markets, where prices had soared, costing its treasury hundreds of millions of euros on top of its fossil fuel subsidies.

Carbon emissions, in the case of crypto miners, are academic: miners don't inherently produce any greenhouse gases because their computers can use power from any source. In a 100 per cent renewable energy system, there wouldn't be any emissions. Bitcoin miners are, in any case, at the vanguard of clean green energy: at least 40 per cent of Bitcoin mining is green today, and 75 per cent of miners already use some green energy and want more. By contrast, in China, renewable energy as a percentage of total energy consumption is 29.5 per cent. The EU is greener, at 34.1 per cent, while the US lags behind, at only 12 per cent. In addition, major global sectors such as steel, cement, aviation and shipping have undergone almost zero greening, and other industries, such as building and construction, waste processing and water desalination, aren't doing much better.

We also need to keep things in perspective. In 2018, the annual electricity consumption of Bitcoin was 45.8 terawatt-hours (TWh), which from a pollution perspective means its annual carbon emissions were 22.9 million metric tons of carbon dioxide. A couple of years later, this had gone up to approximately 110 TWh (about what the Netherlands consumes in electricity per year), creating approximately 53.32 million tons of CO_2 (roughly the greenhouse gas pollution Singapore emits each year, or twice the greenhouse gases video gamers generate).

The electricity consumption of Bitcoin will continue to rise until the last of the 21 million bitcoins is mined in about a decade; around 19 million had been mined by 2022, and the 21 million is a hard cap – no more bitcoins will ever be added. That's because normal computers are no longer able to do the job, which is in part a feature of the system's integrity and security: the more coins that are mined, the more complex is the proof-of-work algorithm required to add blocks to the Bitcoin blockchain, which in turn requires more equipment (literally warehouses filled with computers), better and more expensive processing power and more electricity.

But how much electricity is all this in relative terms? The answer is not much. The world consumes approximately 160,000 TWh per year, and wastes along the way at least a third of this, or approximately 55,000 TWh per year. What Bitcoin consumes (110 TWh per year) is therefore a fraction of 1 per cent of both the world's energy and the wasted energy discarded along the way. 'Always-on' electrical devices in America alone (such as washing and drying machines, dishwashers, televisions, etc.) consume 12 times the electricity of the global Bitcoin mining network.

The competition, gold, is far worse from an energy-intensity perspective. Gold is mined, refined and processed into jewellery, bars and coins, and electronics. This life cycle generates 123 million tons of CO_2 each year, according to the World Gold Council, which equates to approximately 240 TWh per year of electricity consumption, more than double that of Bitcoin.

Bitcoin should never have become such a controversial issue. Bitcoin assets represent just \$1 trillion, or 0.25 per cent, of our \$400 trillion global financial system. What differentiates them is the fact that the blockchain enables an independent, decentralized and digital monetary system where trust between two parties is established more or less instantly using electricity. Bitcoin's energy use is in fact a harbinger of new business models and applications, which will increasingly be based on the coming abundance of near-zero marginal cost and zero-carbon renewable energy.

It's been clear for some time that the energy sector is under-going the most profound transformation it has experienced in over 100 years, rapidly decarbonizing by shutting down and decommissioning coal-powered power plants and natural gas power plants and investing in new capacity that is almost exclusively (over 90 per cent) renewable energy. This is driven by economics, because solar, wind and batteries are cheaper than coal and natural gas even before taking the destructive environmental impact of fossil fuels into account: over the past decade, solar power costs have decreased by 90 per cent, as have those of lithium-ion batteries, while wind power costs have decreased by 50 per cent, with more improvements to come.

As we build more and more solar power plants, wind power plants (both onshore and offshore) and battery storage facilities, their disruptive impact is increasing because of economies of scale and continued reductions in their cost. We are combining solar, wind and batteries in the most cost-effective way possible – this will vary from place to place, depending on a particular location's weather characteristics – by creating very large quantities of solar and wind capacity, offsetting the need for storage requirements. That's because when building an electricity generation system based on solar and wind resources, days with no or low sunshine and others with no or weak winds have to be taken into account. Therefore solar and wind power plants will provide a multiple of the electricity needed on any given day, thus minimizing the size of the batteries required to store energy for dull or non-windy days. This paves the way for an eruption of new business models to capitalize on these vast quantities of clean energy electrons.

Technological disruption isn't linear. Information technology created an almost infinite abundance of binary digits (bits), which replaced and complemented finite physical, tangible and analogue products across multiple industries, such as newspapers, vinyl LPs, cassette tapes, CDs and VHS videos, and along the way created entirely new models at a much larger scale (think Netflix versus Blockbuster), operating on entirely different principles.

The disruption created by bits is a preview of what electrons are currently in the process of doing. Once increasingly massive amounts of excess energy are made available at near-zero marginal cost, entirely new business models and applications will emerge. We can already see the advent of the electron abundance era in the increasing electrification of cars, scooters and buses. This will extend to multiple other sectors, such as water treatment, green cement and green steel, waste processing and many others. Bitcoin miners are just at the vanguard. Erroneously focusing on how much energy they use is a trivial pursuit.

25

Love Thy Insect

If you drive, have you noticed that not so many insects are crashing into your windshield compared to a few years ago? This even has a name now, the 'windshield phenomenon'. To put it simply, we are in the process of wiping out insects. This should be front-page headline news every day. We like to think we run the world, but we are not the world's ultimate controllers. The ultimate controllers, the tiny things that enforce our world order, are insects.

They may be tiny, and you may almost never give them a second thought unless you are swatting them away, but they are mighty and important. They are everywhere on land and they are everywhere below ground. They are everywhere in the sky. They aren't really in the oceans, but some love water: the viruses that cause dengue fever and Zika are carried by the Aedes mosquito, which breeds in clean, stagnant water easily found in some of our homes.

You can't even reliably estimate the number of insects because there are so many of them. They are in the millions of trillions.

Their number is so large that we've never really tried to count them properly, but it's estimated that the world's arthropods, a group that includes insects and spiders, are likely to weigh at least 20 times more than all 7.7 billion human beings combined. They are intimately, intricately involved with everything on earth. They toil on and in every cubic centimetre of living soil. They aerate, they fertilize.

Insects are the recycling champions of the world. They digest dead wood and dead bodies. They break down billions of pieces of organic debris and waste that people and animals produce. Or at least they try to, unless we make it impossible for them with certain kinds of plastic waste. They live at the base of the food chain and so directly or indirectly feed millions of birds, mammals, reptiles, amphibians and fish, and of course, by extension, us. They pollinate more than 300,000 species of plants.

Take away this enormous mass of insects – they may be 90 per cent of all animal species – and you would be looking at planetary ecological breakdown: waste would pile up, the soil would shed its nutrition, animals would starve, and hundreds of thousands of plant species would vanish. Think of your everyday meal consisting of only what is wind-pollinated: you'll still have bread, but you won't have any fruit or vegetables. Most of the meat will also be gone, and you can forget about chocolate as well.

In monetary terms – if it even makes sense to think about ecological breakdown using the concept of money – we're talking hundreds of billions of dollars' worth of food supplies, and even that's hard to compute because of the sheer enormity of the mass of insects that nobody's ever counted: we think more than 75 per cent of our food crops rely at least in part on pollination by insects and other animals.

Take bees, for example, a very important family of insects we've managed to classify and categorize into 20,000 different species. They can be broadly divided into solitary bees (those that don't live in colonies) and social bees (organized in groups, each of which has a queen, male drones whose only role is to mate with an unfertilized queen, and hard-working female worker bees in charge of housekeeping, feeding everybody, and everything related to making honey).

Bees are what's called a keystone species, which means that other species are dependent on them to survive. Their importance and reach are staggering: someone or something has to move pollen from the male part of a plant to the female part for fertilization to follow. That job is overwhelmingly done by animals, including butterflies, bats and moths, except it's mainly the bees that deliver the heavy lifting. In economic terms, bees contribute some $500 billion to the world's economy each year through the global crops that would not be there if it weren't for them.

There's lots of other stuff that bees do that's wonderful. Honey bees have been producing incredibly healthy honey packed with vitamins, minerals and enzymes for 150 million years. It's only recently, with the evolution of human beings, that they have become threatened. The non-honey-producing bumblebees, some of the largest and gentlest of bees, are the pollination champions of the world and deliver the tomatoes and cucumbers many of us consume daily, as well as a wide variety of seed crops and fruit and vegetables, including berries, melons, peppers and squash. Bees also produce wonderful beauty treatments that have been used since Cleopatra's time.

To put the importance of bees into perspective, a supermarket in Hanover, Germany, emptied its shelves of products that depended on bees for pollination. The result was that 60 per cent of the groceries had to be cleared, including apples, courgettes, almonds, coffee, avocados, onions, berries, chocolate and sweets, some marinated meats and even camomile-scented toilet paper.

In a nutshell, without bees the variety of foods available would shrink tremendously and the cost of what's left would surge. We would be left with a lot of calories, but these would come from wind-pollinated cereal grains. For the rest, we would be dependent on massively labour-intensive and expensive manual pollination, and possibly, in time, robotic pollinator drones, if we could ever produce these at reasonable cost and they could mimic what the amazing bees do.

Bees have been under sustained attack by human beings for decades, sometimes knowingly and other times unknowingly. In North America, for example, you are 50 per cent less likely to see bumblebees today compared to before 1974, and in Europe, their numbers have decreased by 17 per cent compared to the early twentieth century. Many bumblebees are therefore listed as endangered, vulnerable or near-threatened. While so-called pests and parasites have plied their trade for millions of years, we have brought them new threats in the form of pesticides and climate change.

However, with the exception of governments going rogue from time to time and green-lighting pesticides that we know kill bees – as happened under the administration of President Bolsonaro in Brazil – we're by and large listening to the science and banning those pesticides we know are harmful to insects and wildlife. As a result, most of the decline in the population of bees worldwide is today driven by climate change.

The south-west of England is a potent case study, having turned into a hotspot for bee diseases as temperatures warm. Honey bees were found to be 39 per cent more likely to catch the fatal Varroa mite disease for every 1° Celsius increase in temperature, while other English areas have seen increased bacterial honey bee diseases also powered by a warming climate.

In addition to food crops, pollinators contribute to crops that provide, for example, palm oil, fibres such as cotton, medicines, food for farm animals and even construction materials. Even though we might not know in any great detail how many insects we have and exactly how many plants they pollinate, we know enough to understand that if they were not around, we would have an ecological disaster of planetary proportions.

Let's get back to the windshield phenomenon. Even though most of us aren't much interested in insects (they're annoying, and when you see one, you basically just want to whack it), at a larger scale we've now officially noticed that there are fewer of them. Until recently, science hadn't done much to track insect decline. Much of the intuition that there were fewer insects came from biodiversity databases, records of species sightings and information gathered by volunteers, but generally it was a piecemeal picture, complemented by our knowledge of the impact of forest fires on animals and insects. One of the few documented studies of the impact of wildfires on wildlife showed that three billion animals were killed or displaced during Australia's devastating bush fires of 2019 and 2020. The numbers are hard to visualize, let alone comprehend: according to the World Wide Fund for Nature and the calculations of a dozen Australian scientists, 143 million mammals, almost 2.5 billion reptiles, 180 million birds and 51 million frogs were killed. The insects wiped out are likely to be multiples of these very large numbers.

There is, however, one study that gives us a sense of the extent of the carnage. A German team gathered a massive biodiversity sample – I can't even get my head around how they did it – of more than a million insects across 2,700 species in 150 grass fields and 140 forests in three regions in Germany. They were at it for 10 years, between 2008 and 2017, which in itself is astonishing.

What they found was terrifying. Several rare species had completely disappeared. The total biomass, the weight of the insects in the forests, was down a massive 40 per cent over the 10-year period. The weight of the insects in grass fields was down a breathtaking 67 per cent. All types of forests and all types of land with grass were affected, whether fertilized or not, whether the farming was organic or not. Finally, insects' abundance in terms of both the number of individuals and the number of species was declining dramatically in both grass fields and forests.

We have a good idea why we're eliminating so many insects: we are destroying nature through changes in land and sea use, through turning intact tropical forests into agricultural land without regard to conservation, biodiversity or sustainability. That's before we even get to pollution from oil, gas and coal and the impact of climate change, which in effect accelerates and worsens everything.

Natural extinctions graduate to become mass extinctions when we lose the majority of our species in a short geological time because of an event like a meteor, a volcanic eruption or an asteroid. We've had five of those extinctions in 600 million years and we're smack bang in the middle of the sixth, except this one is very different: the catastrophic natural event this time is us – humans. We are driving this sixth mass extinction on super-fast-forward: the rate of species extinctions today is somewhere between 100 and 1,000 times faster than the normal rate.

Our best strategy in response to this mass extinction is to stop then reverse deforestation, and we have a powerful tool to do this: set aside half of the earth's surface for all other species and keep 50 per cent for us. Countries can agree among each other how best to allocate this 50 per cent, so that, for example, people who are squeezed into skyscrapers don't have to give up any land, whereas those who have thousands of square metres they're doing very little with donate it as a nature reserve. As an added benefit of planetary significance, the insect apocalypse would also begin to reverse.

The 50 per cent goal, necessary as it might be, remains aspirational. Concrete, systemic steps are, however, being taken to head towards it over time. The most promising initiative, the High Ambition Coalition for Nature and People (co-chaired by France, Costa Rica and the UK), combines 70 countries who have pledged to protect at least 30 per cent of the world's land and oceans by 2030, by law, in order to tackle biodiversity loss and fight back against mass extinctions. This initiative was prompted by a paper written by 19 prominent scientists, who argued that it should be the centrepiece of a proposed Global Deal for Nature.

Thirty per cent was selected because the scientific community is arguing persuasively that the science is clear: at least that proportion of the planet's land and ocean must be protected to address the alarming collapse of the natural world. The 70 countries grouping under the High Ambition Coalition for Nature and People banner are strongly pushing for this '30x30' goal to become enshrined in the 10-year strategy (referred to as the Global Biodiversity Framework) of the United Nations Convention on Biological Diversity, which 196 countries are trying to finalize via annual meetings similar to the annual climate talks.

Both the 30x30 goal and the Global Biodiversity Framework (a first draft of which was released in 2021) are necessary, because for now, the future of insects doesn't look bright: the world's governments continue to spend vastly more on destroying biodiversity than on protecting it. According to the OECD, the organization grouping the world's high-income and developed economies, positive global biodiversity spending – by governments and the private sector combined – is between $78 and $91 billion each year, whereas governments alone spend approximately $500 billion each year on activities harmful to biodiversity. That's five to six times more money for destructive spending even before taking into account spending by the private sector. Yet we know how much we would benefit from protecting our lands and oceans. The benefits are at least five times the costs: protecting nature drives economic growth and saves money. What's there not to like about that?

Encouragingly, philanthropic organizations understand the importance of protecting 30 per cent of the lands and the seas by 2030. Nine prominent ones, for example, pledged $5 billion (spread out over 10 years from 2021) to this effort, in the largest donation to nature conservation in history. The organizations – Arcadia, the Bezos Earth Fund, Bloomberg Philanthropies, the Gordon and Betty Moore Foundation, Nia Tero, the Rainforest Trust, Re:wild, the Wyss Foundation and the Rob and Melani Walton Foundation – are specifically directing their funding to projects supporting the initiative championed by the High Ambition Coalition for Nature and People. More private-sector philanthropy will undoubtedly follow. This, however, is the road less travelled: 'fixing' biodiversity by 2030 requires spending an extra $711 billion on the problem, according to an economic study by the Paulson Institute, the Nature Conservancy and the Cornell Atkinson Center for Sustainability at Cornell University. While the number isn't as daunting as it looks at first glance (it's less than 1 per cent of annual global GDP, or to put it another way, just double what we spend on shoes each year), it is still hundreds of billions of dollars. In the age of coronavirus, we know that governments can muster instantaneous trillions of dollars overnight when they absolutely must. Protecting biodiversity is an absolute must.

While waiting, we could use small acts of thoughtfulness and kindness. If you have a garden, consider rewilding. If you use pesticide, consider cutting its use. If you need outdoor lighting, consider limiting it. Help change attitudes towards insects by conveying, where you can, that they are crucial components of our living world. When you see that little buzzing pest again, think about the insect apocalypse, then get angry.

26

The Royal Baby Versus Biodiversity

In May 2019, an obscure United Nations body, the Intergovernmental Science-Policy Platform on Biodiversity and Ecosystem Services (IPBES), released the 60-page summary of its 1,500-page report on biodiversity. Biodiversity is basically everything: all the variety of life that can be found on earth (plants, animals, fungi, micro-organisms and us), as well as the communities they form and the habitats in which they live. The IPBES summary was a clarion call. It said that because of us, nature is declining globally at rates that are unprecedented in human history. It also said that the rate of species extinctions is accelerating, with grave impacts on people around the world. Many in the biodiversity field were fascinated as they pored over the findings. Yet for the following 48 hours, the IPBES report received almost zero coverage in the media.

The English-language news was dominated by a new royal baby in Britain, President Trump having another go at China, Turkey deciding to rerun elections in Istanbul, two journalists released in Myanmar, the Australian prime minister being hit by an egg and an oil company proposing to take over another oil company to prevent a third oil company from doing the same. It wasn't just the English-language media. There was no coverage either in the French, Spanish, Arabic, Filipino, Singaporean, Chinese or Thai media.

The IPBES may have a mouthful of a name, but it is very important. It is a body set up by 94 governments and administered by the UN Environment Programme. Its report was compiled by 145 expert authors from 50 different countries who had been working for three years to put it together. They'd also gathered input from another 310 contributing authors and reviewed 15,000 sources. The summary report made for extremely sombre reading.

It documented the fact that 25 per cent of plants and animals are threatened, with a million species already facing extinction, and warned that the global rate of species extinction continues to accelerate from a level that is already tens to hundreds of times higher than it has averaged over the past 10 million years. Of the 6,190 domesticated breeds of mammals we use for food and agriculture, over 9 per cent are extinct while at least another 1,000 will be soon. Fish aren't spared either. Marine plastic pollution is up 10 times since 1980, while 300 to 400 million tons of heavy metals, solvents, toxic sludge and other waste from industrial facilities is dumped annually into the oceans. Fertilizers have produced 400 ocean dead zones, an area greater than Britain. This is affecting 86 per cent of marine turtles, 44 per cent of seabirds and 43 per cent of marine mammals.

What is it about the media that means it invariably ignores biodiversity loss and extinction risks and instead chooses to focus on stories like a royal baby? Professionals in the mainstream media are generally informed, diligent, hard-working and smart. Surely they are not ignoring biodiversity because they don't know about it. It must be because the royal baby sells papers and attracts clicks, whereas the collapse of biodiversity doesn't.

The plot thickens when looking at climate change coverage in the media, or the relative coverage of climate change versus biodiversity. Media coverage of climate change is eight times higher than that of biodiversity, yet biodiversity loss also threatens our planet's integrity and balance. Biodiversity science and challenges in particular aren't reaching the public at large, even though the IPBES has been releasing facts and reports – backed by extensive presence on social media – since 2013.

One challenge, perhaps, is that the Intergovernmental Panel on Climate Change (IPCC), the UN body for assessing the science related to climate change, was established 25 years before its cousin the IPBES. IPCC reports have therefore been socialized for much longer. Studies indicate that it took approximately 10–15 years for the media to begin noticing the IPCC's climate change warnings and analyses. Perversely, attacks on climate science helped improve the visibility of the reports, because controversy sells. The media also loves catastrophic wildfires, typhoons, heatwaves and floods, and it was hard pressed to ignore the increase in their incidence.

The Media and Climate Change Observatory, a multi-university collaboration, monitors 126 sources across newspapers, radio and TV in 58 countries around the world and publishes regular reviews of media coverage of climate change. Its data indicates that even in relation to climate change, media attention has been inconsistent: we are still to match the heights in coverage of December 2009, when world leaders converged for the UN's fifteenth year of climate talks in Copenhagen. 2021 is in second place.

Biodiversity only becomes obvious as a critical global challenge when its manifold local manifestations are aggregated. It's therefore more challenging to track within a news cycle focused on short-term events such as a royal baby. Readers can help by directing their money towards media that's reporting the really important stuff. If all we want to read is about the royal baby, then that's what we will get.

COVID-19 should be a thunderous wake-up call. In 2008, a team of researchers from the Zoological Society of London, the Wildlife Trust and Columbia University identified 335 new diseases that emerged between 1960 and 2004, 60 per cent of which spilled over from animals, just like COVID-19, MERS and SARS. Warnings to expect more deadly viruses were abundant (and still are), but aren't heard loudly enough.

We're seeing more coronaviruses infecting humans in the last 50 years because of human activity. Deforestation, air pollution, illegal wildlife trade, intensive and unsustainable agricultural practices and industrial fishing are wrapped into a climate change crisis – a threat multiplier – and create a tornado of biodiversity loss. Wetland – land flooded by water, sometimes seasonally, characterized by its vegetation of aquatic plants – starkly illustrates our impact: just 13 per cent remains compared to what there was in 1700. In turn, pathogens leave the wild, and virus spill-over risks from wildlife to people rise as contact between us and them increases.

In addition, virus bombs (multiple viruses appearing simultaneously) are inevitable if we don't act to protect biodiversity by fighting back with far more urgency against climate-change-driven permafrost thawing.

Permafrost is any type of ground that's been frozen continuously for hundreds to thousands of years. It extends down beneath the surface of the earth to sometimes more than a couple of kilometres. Permafrost represents 25 per cent of the northern hemisphere and is widespread in the Arctic regions of Siberia, Canada, Greenland and Alaska. When plants and animals die, microbes from their decomposed bodies release all sorts of gases, including greenhouse gases. The permafrost puts a deep freeze on that process, preserving organisms and gases that otherwise would have gone into the atmosphere. We have, stored in the permafrost, many thousands of years of life, from humans to plants to mammoths. It's one of our greatest stores of gases and a hidden carbon bomb: in the Arctic alone there's twice as much carbon as is currently in the atmosphere. There's also an enormous amount of methane, another greenhouse gas, which traps 84 to 86 times more heat than carbon over a 20 year period.

It's also a vault of microbes, many of which are ancient. We've already discovered some microbes in thawed permafrost that are more than 400,000 years old. It's quite likely that humans would have no idea how to deal with many of these if they were to resurface. For now, they only do so occasionally. In 2016, for example, there was an outbreak of anthrax in Siberia linked to a decades-old reindeer carcass infected with this bacterium then exposed when the permafrost melted. It is also clear that developing the Arctic – which in some cases means extracting permafrost to mine for precious metals and petroleum – as well as exacerbating climate change increasingly puts permafrost in danger and melts it. Nonetheless, that's precisely what we are doing.

Arctic states and many big corporations are rushing to exploit the area, their enthusiasm increasing proportionately as it melts and its resources become easier to reach and extract. The US government estimates that the Arctic holds slightly less than a third of the world's undiscovered natural gas, and 13 per cent of undiscovered oil, for example. Several countries, including Russia and Norway, together with Big Oil, would like nothing more than to embark on drilling for these oil and gas resources and to build the infrastructure – ports, pipelines, roads – to ship them to market. That's now significantly easier, because melting ice means new shipping routes are opening via the Arctic. The growth of shipping on the Northern Sea Route (an alternative for Europe-to-Asia shipments that avoids the Suez Canal and is significantly shorter) has been explosive over the past few years: in 2022, year-on-year tonnage was up 15 times compared to a decade ago.

The Arctic is already warming twice as fast as the rest of the world, and this ecological chaos has become so stark that we created a new name for it: Arctic Amplification. The area's disintegration is starkly illustrated by new high temperature records across Siberia, as well as unprecedented Arctic wildfires releasing more carbon dioxide than ever. In both 2019 and 2020, the US and Canada recorded smoke from Siberian wildfires that spread for thousands of kilometres, releasing record climate-warming gases. This is increasing the probability that the permafrost will thaw, releasing more carbon bombs as well as an unknown quantity of microbes we are not adapted to. What's required is nothing less than a

Paris Agreement-style biodiversity treaty to stop the carnage, complementing carbon restrictions aiming to stop emissions-heavy investments.

Meanwhile, the failure to address biodiversity is stunning. In 2010, 190 countries adopted a Strategic Plan for Biodiversity 2011–2020 under the auspices of the United Nations, while the entire decade was named the United Nations Decade on Biodiversity. Twenty goals were set. None were achieved, and many were missed in a particularly spectacular fashion. For example, we had such lofty aims as 'by 2020, the rate of loss of all natural habitats, including forests, is at least halved' and 'by 2020, all fish and invertebrate stocks and aquatic plants are managed and harvested sustainably'. But deforestation continues unabated and industrial fishing has worsened since 2010.

Some aims, in particular the designation of significant land and ocean areas as protected, should be prioritized. The 2020 target set in 2010 was for at least 17 per cent of lands and 10 per cent of coastal areas to be protected, a goal also missed. COVID-19, SARS and MERS may be just the tip of the iceberg unless biodiversity is protected and promoted.

27

Sue the Bastards

Within the space of four weeks in 2021, Germany's constitutional court ordered the German government to significantly accelerate the country's action on climate change, while a Dutch court ordered oil company Shell to reduce its climate-warming emissions by 45 per cent by 2030 and said the company must take responsibility for its own carbon emissions as well as those of its suppliers.

The German government obliged, whereas Shell appealed (its appeal is in progress). These two decisions were monumental at the time they were issued. No country had been ordered by its courts to comprehensively step up on climate change in this way. No oil company had been ordered by any court to shrink by almost half. They were but the tip of the global climate litigation iceberg that has boomed since the 2015 Paris Agreement to reach 1,550 cases globally in 2021 (as discussed in Chapter 22).

Climate change cases come in multiple flavours. Some are filed against listed companies, accusing them of selling physical or financial products (for example, petrol, or stocks and bonds) without adequately disclosing the climate risks attached to these, or disclosing them in a misleading way (greenwashing). After one such case in the UK, BP had to withdraw an advertising campaign with the tagline 'Possibilities Everywhere' because it was misleading the public about the scale of its green portfolio and playing down how dirty natural gas was. Shell too was rebuffed when the Netherlands' advertising watchdog, responding to a complaint from a group of nine students, ruled that advertising where the company claimed that customers could offset the carbon emissions from their fuel purchases was misleading.

Another type of climate lawsuit involves going after polluters for violating people's constitutional rights, such as the rights to life or health. In the 2021 Shell case, the Hague District Court ruled that the Dutch government wasn't taking sufficient action to protect the right to life. It ordered Shell to improve its emissions in a manner consistent with limiting warming to less than 1.5°C compared to pre-industrial times, finding that every CO_2 emission by the company anywhere in the world contributed to climate change in the Netherlands – potentially endangering the right to life – and hence Dutch law applied. Over in Portugal in 2020, six children filed a lawsuit in the European Court of Human Rights against 33 European countries for threatening their right to life by moving too slowly to reduce greenhouse gas pollution. This case is still being litigated.

A third type of case seeks to force countries and oil, gas and coal companies to keep fossil fuels in the ground. In Pakistan, a seven-year-old girl represented by her father sued to stop particularly dirty coal fields from being developed, on the grounds that these mining activities would drastically increase Pakistan's emissions and be catastrophic for future generations. In Colombia, the Supreme Court ruled in 2018 in favour of 25 young plaintiffs seeking to ban mining and oil exploitation – effectively blocking 473 existing concessions – in the country's páramo, the delicate ecosystem nestled between the forest and the permanent snowline. In Kenya, the National Environmental Tribunal invalidated a licence granted for a new coal-fired power station.

Another flavour of lawsuit is concerned with the enforcement of legislation or regulation already on the books of national and subnational governments, challenging them if their commitments are not being put into practice. The French government, which brokered the landmark 2015 Paris Agreement, was hauled before the Council of State by Grande-Synthe, a low-lying northern coastal town that is particularly exposed to the effects of climate change. The Council, which rules on disputes over public policies, reminded the government it was failing in its commitment to reduce emissions by 40 per cent come 2030 compared to 1990 levels. Before issuing a final ruling, it gave the government three months to justify how its refusal to take additional measures was consistent with its own targets.

But what all these actions have in common is that they seek to hold power – governments and polluters – to account. Out of the approximately 1,550 cases worldwide in 2021, the overwhelming majority, 1,200 (77 per cent) are in the US. However, they are spreading rapidly elsewhere. Number two was Australia, with approximately 100 lawsuits, followed by the UK at 58 and the EU at 55. There are many more in almost every EU country as well as Ukraine (though Russia is conspicuously absent in Europe, as is China in Asia), and more are being added constantly.

No continent is spared. Climate lawsuits are everywhere in Latin America, including Argentina, Brazil, Colombia, Chile, Ecuador, Mexico and Peru. It's the same story in Asia (India, Indonesia, Japan, Pakistan, the Philippines, South Korea and Taiwan), though Africa is somewhat under-represented, with cases just in Kenya, Nigeria, South Africa and Uganda.

Climate lawsuits are not about whether we should use fossil fuels. Reasonable observers accept that we will never give up on fossil fuels entirely; they will still be needed for essential goods we produce using oil and gas. But these lawsuits matter because of the key questions they're trying to answer, ranging from who should pay for damages from fossil fuels to who should pay for the fact that we've known for over 40 years that increased use of oil, gas and coal is destructive to our health and to that of the planet.

The lawsuits terrify oil, gas and coal companies, because it is a matter of public record that they've known about the harm their products caused in the same way that tobacco companies did about smoking. After all, oil companies have been building defences against sea level rise on their offshore oil rigs and oil refineries for decades.

Imagine what would happen if an oil executive from ExxonMobil testified in open court that they had knowingly misled the public about the climate threat since at least 1980. That's exactly what happened in the case of the tobacco industry. The fossil fuel industry would then be liable to pay extremely large fines and compensation, which could potentially mean bankruptcy.

In 1988, the tobacco companies signed up to the largest civil litigation settlement in US history, which forced them to stop advertising, limit lobbying, restrict product placement, fund anti-smoking campaigns and pay more than $200 billion in compensation over 25 years. The oil and gas industry is potentially on the hook to pay a hundred times that much, which is going to be necessary for communities everywhere to fight back, repair the damage and adapt to the climate catastrophe.

Oil and gas fines will be greater than those of the tobacco industry because the sector is far more destructive: it has fundamentally harmed the fabric of our planet and our civilization in the name of short-term profits. In addition, while tobacco harmed smokers and those around them affected by passive smoking, greenhouse gases from fossil fuels harm everyone.

In time, climate change litigation will create large financial liabilities for oil, gas and coal companies as well as many other companies that use a lot of fossil fuels in their products, such as petrochemical, cement and steel concerns. The lawsuits also have the potential to suddenly accelerate the shift to green economies: as mentioned above, just one lawsuit achieved that goal in Germany, when, siding with young climate activists, Germany's highest court ordered the country's government to significantly accelerate climate action, paving the way for entirely new businesses and business models across every sector of the economy, all of which are required fast by law.

Climate change lawsuits could also lead governments to treat climate change as a health problem, not just an environmental one. Picture this health warning on the side of diesel-powered buses and cars, plastic products, gas stations, ships, planes and manufacturing plants: *Use of this product leads to a serious threat to the economic well-being, public health, natural resources and environment of everyone on earth through increased climate change impacts – including lethal pollution; loss of sea ice; accelerated sea-level rise; longer, more intense heatwaves, wildfires and droughts; stronger and more intense hurricanes and typhoons; and accelerating species extinctions around the world.*

The direction of travel is also clear, with litigation expanding to fiduciaries, that is people managing money for others, who are not pricing climate risks in their investment decisions, and financial institutions, who back fossil fuel projects and make these emissions a reality.

The wheels of justice may be slow, but the courts will increasingly scrutinize the actions and inactions of institutional investors, fund managers, pension providers and trustees. Directors of public companies are also starting to feel the heat, with efforts under way to hold them personally liable for environmental destruction. For example, in a world first, ClientEarth, the not-for-profit group, notified Shell in 2022 that it would commence legal proceedings against the company's 13 executive and non-executive directors in UK courts over the mismanagement of climate risk and the board's failure to adopt a strategy truly aligned with the 2015 Paris Agreement. ClientEarth will argue that the longer the Shell board delays implementing such a strategy, the more likely it is that the company will have to execute an abrupt handbrake turn to retain commercial competitiveness.

Emerging trends also include potentially more companies being delisted by stock exchanges under pressure from climate lawsuits arguing that what the companies do is destructive to the environment and that exchanges listing such companies share in their liability.

In the climate change space, what is radical has a track record of becoming normal in very short order. It's hard to remember the days when Greta Thunberg was an obscure Swedish student, Alexandria Ocasio-Cortez was a little-known activist or Extinction Rebellion had yet to disrupt commutes around the world. But that has all happened since 2019. Germany's climate plan for net zero by 2045 shows how cost-effective decarbonization can be immediately implemented, on a large scale, once the political will is there.

See also:
Chapter 12: Have as Many Babies as You Like;
Chapter 22: Tinker, Lawyer, Banker, Fry

28

It's Raining Renewable Energy

It's official: in 2020, the International Energy Agency, the Paris-based autonomous intergovernmental organization that acts as one of the world's foremost energy authorities, declared solar power the cheapest source of electricity in history. Wind power wasn't far behind. In other words, both solar and wind power are now below what it costs to build new coal-fired or gas-fired power plants, and can even compete with existing coal plants, which means that rationally, no new coal-fired or gas-fired plants should be built and existing ones will soon be consigned to the power industry garbage bin.

Partly because the corporate world had already figured this out, renewable energy announcements have been coming thick and fast, and will continue to do so for many years to come. Oil companies in particular are trying to jump to the head of the queue. Total, the French oil and gas giant, put renewable energy at the heart of its revamped strategy and rebranded along the way as TotalEnergies. It's planning to invest €60 billion over 10 years in renewables and says it aims to be among the top five renewable energy businesses by 2030. To achieve that goal, it is seeking to deploy 10 gigawatts of renewable energy per year, three times South Africa's existing solar capacity, or the equivalent of France's entire solar energy base.

Shell says it wants to eliminate 40 per cent of its oil and gas costs – tens of billions of dollars – to free up cash to invest in clean energy. It wants to deliver renewable energy to the equivalent of 50 million households, offer electric charging infrastructure to cars and buses and develop a green hydrogen business. It too repositioned its brand, presenting itself as 'global energy company Shell'.

Oil major BP is also on message. It says that renewable energy will be the fastest-growing energy source anywhere over the next 30 years and is on record acknowledging that clean, green wind and solar power are now suitable for what it refers to as newly electrified sectors such as transport, heating and buildings. As a result, it tells us that it's embarking on a journey to become a clean energy giant too and wants to approve 20 gigawatts of renewable energy by 2025, raising its previous target eight times.

American oil majors are getting there as well, though more slowly. ExxonMobil was in shock when a climate-focused shareholder activist won a bid to install new members on its board of directors, against its wishes, a setback it had never encountered before in its 150-year history. The company was planning to double down on oil and gas production, even in the face of a climate catastrophe. On the same day, shareholders overrode the Chevron board to insist the company reduce its emissions from fossil fuels.

The same picture emerges when looking at oil commodity traders. These are relatively obscure names, but very large businesses. Just the top 10 global oil trading companies have $2 trillion in annual revenues (about the size of the Canadian economy) and employ over half a million people.

The largest one, Vitol, is privately held and boasts revenues of $230 billion, more than the size of the economy of Greece or Portugal. Vitol is the world's largest trader of physical oil and gas, the business of moving oil from point to point and managing all the associated logistics. You need a tanker with jet fuel, liquid natural gas or unrefined crude oil? Vitol is generally your first port of call. The second largest trader of physical oil and gas is Glencore, with revenues of over $200 billion. It's listed, and is also one of the largest traders of metals and minerals in the world. The third and fourth, Trafigura and Mercuria, with revenues of $200 billion and $120 billion respectively, are privately held.

Most of the time, none of us have any idea what Vitol, Glencore, Trafigura or Mercuria are doing (they're all very secretive), except that they are trading an enormous amount of oil and that three quarters of a trillion dollars flow into their coffers every year. That's about the annual GDP of Saudi Arabia or Switzerland. We now also know that they want to flex their financial muscles in the renewable energy sector. Trafigura launched a renewable energy business to invest in solar, wind and storage projects and wants to own 2 gigawatts in the next couple of years. Vitol is already doing the same thing. Mercuria is putting $1.5 billion into battery storage projects in the US.

Dwarfing these announcements, China pledged unconditionally to cut emissions to net zero by 2060, something it has never done before in decades of climate change negotiations. To achieve that goal, it stated that it will triple its solar deployment to 105 gigawatts a year (that's 8 times the cumulative installed capacity of the UK) and double wind construction to 50 gigawatts a year. That's China alone adding each year more solar and wind power plants than the entire world did as recently as 2018.

These are incredibly large numbers and indicative of the wall of money – $350 billion in one year – flooding into renewable energy following the net zero commitments of China, the European Union, Japan, South Korea, the UK and the US, which together account for 56 per cent of global carbon emissions and 75 per cent of the world's economy.

But what does this all mean?

Expect the growth of offshore wind to go through the roof. The US, Europe, China, Japan, Taiwan and Korea are already witnessing an acceleration in offshore wind deployment. This type of renewable energy plays to the strengths of oil and gas companies, who have been building offshore oil and gas exploration projects for decades.

Offshore wind energy is generated from the wind blowing over the high seas, which is higher and more constant than over land areas because of the absence of any barriers. Mega-structures are installed on the seabed, in depths of up to around 50 metres. Floating turbine technologies also allow installations in much deeper areas of the oceans, up to one kilometre.

A window into the future growth of this segment was provided by the UK, which announced in 2020 that its offshore wind infrastructure would be capable of powering every home in the country by 2030. By implication, this means about $70 billion of investment is on the way to increase the country's offshore wind power four-fold, to 40 gigawatts. This type of explosive growth has solid precedent: offshore wind capacity in the UK has already increased by 10 times in 10 years, while costs decreased by two-thirds. Costs will continue to decline as capacity expands at a rapid clip. In January 2022, Scotland announced a historic offshore wind auction of 25 gigawatts of offshore wind project development rights, almost single-handedly delivering on the UK government's 2030 pledge.

Solar energy too continues its unstoppable march forward to world energy domination. According to the IEA, renewables – led by the 'new king', solar – will deliver over 80 per cent of new electricity by 2030 and overtake coal as the world's number one source of electricity by 2025.

The massive infusion of liquidity to make all this happen is accelerating deployment in China, India, North America and Europe, with other regions such as Latin America, South East Asia, Central Asia, Africa and the Middle East left with no choice but to try and keep up: economies decarbonizing fast will erect carbon frontiers through the adoption of carbon border taxes.

The EU – whose goal is to reduce emissions by 55 per cent by 2030, one of the most ambitious emissions cuts programmes in the world – is leading the way in showing how this will work. It wants to implement carbon border taxes on imported goods in order to protect businesses within its borders against cheaper imports from countries with less ambitious climate policies. In addition, because emissions from imported goods account for at least 25 per cent of the emissions from goods consumed or processed in the EU, reducing them is critical to any effort to reduce the EU's own emissions. The carbon border levy would therefore be calculated to equalize carbon pricing inside and outside the EU, whether or not exporting countries price their pollution, while exempting goods from countries compatible with the EU's approach. The very same carbon border taxes are under consideration in the UK and the US as well.

For example, steel produced using fossil fuels in countries with no price on carbon emissions, such as Japan, would attract a carbon border tax when it reaches the EU's border. The tax would be calculated on the emissions from that steel – for example, $50 per metric ton of CO_2 emissions – to level the playing field between a producer manufacturing steel in the EU and one manufacturing steel in Japan. Carbon border taxes implemented by large economies such as the US, the EU and the UK will influence trading flows while accelerating renewable energy deployment further. European importers might then favour Australian steel produced using green hydrogen over Ukrainian steel produced with less carbon-efficient methods, pushing Ukraine to accelerate its renewable energy deployment.

Investments in renewable energy might be too little, too late for oil companies facing potential major losses on their existing oil, gas and petrochemical assets. For everyone else, however, a virtuous cycle is kicking off, one where more money means more renewable energy deployment, leading to the building of larger manufacturing capacities, which in turn drives costs further down – for solar, wind, batteries, electric cars, electric buses, electric heaters and so on – and thus results in more deployment.

The energy transition to lifestyles powered by clean, renewable energy isn't a necessary sacrifice, it's a necessary bonus. A ground-breaking 2021 Oxford University paper overturned the common thinking that decarbonizing will be hugely expensive (in fact it

268

saves us \$14–\$26 trillion), and also showed that renewable energy can, and likely will, displace fossil fuels within 25 years.

The Oxford researchers reviewed 140 years of oil, gas and coal price data to remind us that while the prices of fossil fuels are volatile (as starkly illustrated during the Ukraine war in 2022), after adjusting for inflation they are now very similar to what they were 140 years ago, and there is no obvious long-range trend. They therefore don't follow any cost improvement curves, despite the ghastly amounts of subsidies we – society – provide to producers and consumers of fossil fuels. Consumer of oil, gas or coal aren't therefore paying less, even after 140 years. Nonetheless, for decades, politicians and policymakers have worked on the basis that a clean-energy future is expensive and will involve taking on costs that are much greater than simply continuing to burn oil, gas and coal.

It turns out that deploying renewable energy is cheaper, and executing this deployment faster is cheaper still. The Oxford paper shows how for several decades the costs of solar, wind and batteries have dropped exponentially at a rate of approximately 10 per cent per year; solar power is now 2,000 times cheaper compared to when it first entered commercial use in 1958. This is similar, as a trend, to what the semiconductor pioneer Gordon Moore observed in 1965 in what is now called Moore's Law: that computing would dramatically increase in power, and decrease in relative cost, at an exponential pace. This became the golden rule for the electronics industry, allowing it to make ever faster, more powerful, smaller and cheaper transistors that drive our modern computers, video games, cars and toys. Computer chips in the 1970s only had a few thousand transistors, while today we are packing billions into them, using transistors that are the size of a tiny fraction of a human hair.

The Oxford researchers go on to persuasively and analytically show that if we manage to continue deploying solar, wind, batteries and hydrogen electrolysers on their current exponentially increasing trend, we can achieve a near-net-zero emissions energy system within just 25 years. In contrast, a slower transition, involving deployment growth trends that are lower than current rates, would be more expensive.

Most remarkably, perhaps, the Oxford researchers reviewed the projections of 2,095 major energy models used around the world to predict future renewable energy deployment, and found that on average, the cost of solar power dropped at a rate almost six times faster than forecast by these energy models. Notably, the International Energy Agency failed to forecast the speed of decline of solar costs 20 years in a row. That's because these multiple energy models accounted for neither the effect of Moore's Law nor for that of another stalwart law of the pace of technological progress, Wright's Law, which holds that as cumulative production increases in a given industry, costs will fall by a constant percentage because of production efficiencies.

The cumulative effect of 2,095 energy models missing the solar revolution by a mile is today's widespread misperception that a zero-carbon world costs more than one fuelled by oil, gas and coal, when in fact it's cheaper. In addition, these flawed models have sadly played a central role in guiding energy investments and climate policy in almost every country in the world. As a result, countries underinvested in zero-carbon technologies, and worse still, locked in expensive and polluting fossil fuel infrastructure, which is more expensive to maintain and run than clean energy alternatives.

In reality, once technologies start moving on Wright's Law curves, their cost declines become predictable, to the extent that it's already clear that the faster we continue to deploy zero-carbon technologies, the cheaper they will get.

These cost declines can be accelerated further with supportive policies such as ones dismantling soft barriers to renewable energy deployment. A policy as simple as speeding the permit process of renewable energy projects would significantly impact how much solar and wind power gets built. Another example of decisive transition policies is the UK's commitment to install 40 gigawatts of offshore wind capacity by 2030 (enough to meet half of the UK's electricity needs), quadrupling current capacity and driving cost reductions further. Many more similar policies are needed in order to drive a faster deployment of zero-carbon technologies (and of the smarter grid we require to maximize the impact of zero-carbon power), while increasing the breadth and resilience of their supply chains and of the associated electric vehicle charging infrastructure.

The signs are encouraging. Solar generation was up 23 per cent globally in 2021, while wind power gained 14 per cent. Together, wind and solar energy accounted for 10.3 per cent of total global electricity generation, up 1 per cent from 2020 and doubling their share compared to 2015, when the Paris Agreement was signed. Fifty countries now generate more than 10 per cent of their electricity from the sun and the wind, out of which seven are new in 2021: China, Japan, Mongolia, Vietnam, Argentina, Hungary and El Salvador.

If we can carry through the 10-year average growth rate of 20 per cent all the way to 2030, solar and wind power will account for over 50 per cent of total global electricity generation by then, increasing our chances of limiting warming to 2° Celsius over pre-industrial times.

The 2021 US infrastructure plan is indicative of the fact that more decisive policies are on the way. When enacted, the bill was the largest climate investment in US history, with $150 billion for clean energy and climate adaptation, $73 billion to modernize the electricity grid, $7.5 billion for a nationwide network of electric vehicle chargers, and other goodies.

These decisive policies in turn allow faster progress along the Wright's Law curves. The Oxford paper ran some numbers to find that decisively transitioning to zero carbon would in fact save $26 trillion in energy costs worldwide by 2050, versus the energy system we have today, which continues to be overwhelmingly dependent on fossil fuels. In addition, we would create millions of new jobs while fighting back against climate change, outcomes not included in the $26 trillion win.

Oil companies and oil traders are announcing grandiose renewable energy expansion plans because they have to, having seen the writing on the wall of decarbonization commitments by most major economies.

As I also see regularly in the executive recruitment area, many of their people aren't happy to work at fossil fuel companies, even if most stay because of their large pay packages: they want their employers to do the right thing and move to decarbonize their businesses. The fossil fuel industry knows how difficult it is becoming to recruit young talent, and that's because most of the six million people directly employed by the petroleum industry don't want it to be around at scale any more.

Listen to the talent, see the future.

See also:
Chapter 9: Stinky Gas

Epilogue

Taiyuan, China's coal capital and the city where my climate change journey started with indelible images of bad skin and teeth, has metamorphosed in 15 years into a near-spotless metropolis. In 2016, it became the first city in the world to electrify its entire, 8,000-strong taxi fleet. Every public bus and almost every scooter in the city turned electric. Diesel trucks virtually disappeared, as did household boilers that used coal. The air was cleaner, the sky bluer. By 2021, the city boasted a 'Taiyuan City Winning the Blue-Sky Defense Battle Decisive Battle Plan' initiative, modelled on China's 2018 'Three-Year Plan on Successful Defense of Blue Sky', which set targets to improve the air that citizens breathe.

Taiyuan's citizens had neither switched to plant-based food nor given up flying. They weren't wearing organic clothes either. Most aren't (yet) driving electric cars. Instead, they acted in droves to force authorities to address environmental problems, expressing their wishes loudly and boldly.

Authoritarian China has witnessed tens of thousands of environmental protests a year since at least 2005, when 74,000 demonstrations were recorded, up from approximately 10,000 a decade earlier. In 2016, the film *Plastic China* was released, documenting the lives of two families who derived their income from recycling plastic waste imported from rich countries. Blocked in China, the film went viral nonetheless, and soon afterwards, the Chinese government banned imports of most foreign plastic waste.

The film exacerbated the fury of China's citizens over stifling pollution, and they made their voices heard through frequent demonstrations, which sometimes spilled into all-out riots. Their target was polluting industries – coal and industrial plants – blamed for the environmental degradation. The government had no choice but to respond. Cleaning up pollution became a top-level priority for the communist party, who were fearful of an erosion of support. In 2013, China's president, Xi Jinping, promised to build an 'ecological civilization', and in 2018, he unveiled a superagency to turn the tide on pollution.

The new Ministry of Ecological Environment was charged with curbing greenhouse gas emissions, protecting China's waterways from pollution and delivering cleaner air. Consolidating formerly scattered regulators spread across several ministries, the superagency benefited from its elevated status to fulfil the top-level mandate emanating from China's leader himself. The work taking place to deliver on Xi Jinping's promise is enormous: 80,000 environmental public interest cases were brought by Chinese prosecutors across the country in 2020 alone, seeking to enforce environmental regulations. By 2021, China had trained

1,200 judges and prosecutors, and 16 environmental laws and regulations were drafted.

The action in China is not nearly enough, but it shows that we can – and will – win the fight against climate change. But we need the pressure to be maintained and escalated.

By the end of 2021, governments had spent $17 trillion, according to the International Monetary Fund, to mitigate the effects of the coronavirus pandemic, deploying massive resources to support businesses and individuals and to channel money to shore up economies. In comparison, just $1.3 trillion was spent on climate change by governments and the private sector combined over the same period, or 7.4 per cent of COVID-19 spending by governments alone. The numbers tell the story. First, the world can mobilize quickly and at scale in a crisis. Second, the world 'gets' global short-term crises but has a harder time combating a threat that most people can't tangibly experience on a regular basis. Third, we have the financial resources to decarbonize fast: $17 trillion would get the job of driving emissions of greenhouse gases to net zero by 2050 substantially done. We just have to decide to do it.

The clock is ticking. We need to keep the struggle for a habitable planet focused on consequential levers of change. One of these is the law. Once something becomes law, it becomes enforceable. Then it becomes normal. That's the case in China and Russia just as much as it is in Western democracies. In the twentieth century, great societal changes occurred once the law changed. In many cases, these legal reforms were preceded by popular mobilization. Whether it is women's right to vote, or same-sex marriage, street protests and concerted action over many years, if not decades, is what leads to changes in the law.

Another lever is through mobilizing groups of people who can compel governments, banks, major corporations and the capital markets to act. The groups can be small (just a few individuals) or large (a protest movement). Either way, they can effect rapid change.

Keep putting pressure on corporations and pension funds to do the right thing: write to your pension plan and tell them you want your money in responsible funds and not funds loaded with fossil fuels. Under sustained pressure from hundreds of thousands of citizens, they will have no choice but to direct asset managers and banks to get their act together and move decisively on climate action.

If you work at a law firm, an insurance company or a bank, start with the cafeteria. Gather like-minded colleagues and insist on banning plastic there. Then escalate. Ask for the right to vet new clients. Apply a climate change lens to this vetting. Small circles of influencers have disproportionate leverage. Five individuals at a bank are a protest movement too. Weakening the ability of polluters to tap legal services from the best law firms, financial services from the largest banks or insurance policies from the major insurers weakens their destructive lobbying and reckless focus on expanding their operations.

Larger groups of people, such as the divestment movement, or organizations like Extinction Rebellion – the non-violent protest movement using civil disobedience techniques to compel government action on climate – have proven their effectiveness. We need many more of these citizen action circles. New ones could leverage the blockchain to mobilize capital to directly accelerate the deployment of renewable energy; or to send a wall of money to manufacturers of plastic substitutes, for example.

Individual action such as not using plastic or not eating meat can play a role in motivating others, and it has a feel-good factor attached that matters. It is not, however, going to get us where we need to go. We must resist oil, gas and coal companies trying to shift the burden for solving the climate crisis to individuals. Instead, we must compel them to assume their immensely larger responsibility.

Sources

Introduction

https://unfccc.int/process-and-meetings/the-paris-agreement/the-paris-agreement

https://yaleclimateconnections.org/2021/08/1-5-or-2-degrees-celsius-of-additional-global-warming-does-it-make-a-difference/

https://www.imf.org/en/Publications/WP/Issues/2021/09/23/Still-Not-Getting-Energy-Prices-Right-A-Global-and-Country-Update-of-Fossil-Fuel-Subsidies-466004

https://gizmodo.com/netherlands-officials-tell-shell-to-stop-its-ads-greenw-1847613583

1. Plastic Is Your New Diet

https://pubs.acs.org/doi/10.1021/acs.est.9b01517#

https://www.theguardian.com/environment/2017/sep/06/plastic-fibres-found-tap-water-around-world-study-reveals

https://www.downtoearth.org.in/news/water/even-bottled-water-can-have-microplastics-58676

https://www.theguardian.com/environment/2018/mar/15/microplastics-found-in-more-than-90-of-bottled-water-study-says

https://www.openaccessgovernment.org/microplastics-babies/128723/

https://www.openaccessgovernment.org/babies-microplastics/121533/

https://www.nap.edu/resource/other/dels/plastics-in-the-ocean/

UN treaty: https://www.deeperblue.com/u-n-environment-assembly-adopts-a-plastic-pollution-treaty-resolution/

https://www.banktrack.org/campaign/banks_and_fracked_oil_and_gas

https://www.recyclingtoday.com/article/aluminum-cans-recycled-twice-plastic-bottles/

2. Who Put Palm Oil in My Toothpaste?

https://ourworldindata.org/palm-oil
https://forestsandfinance.org/news/
    banking-centers-of-brazil-indonesia-
    china-the-united-states-and-japan-
    are-bankrolling-global-deforestation/
https://www.gibsondunn.com/
    mandatory-corporate-human-rights-
    due-diligence-what-now-and-what-
    next-an-international-perspective/
https://clsbluesky.law.columbia.edu/
    2021/02/15/gibson-dunn-discusses-
    eu-developments-in-corporate-
    human-rights-due-diligence/
https://www.tcfdhub.org/wp-content/
    uploads/2021/06/Primer_on_Climate_
    Change_Directors_Duties_and_
    Disclosure_Obligations_CGI_CCLI.
    pdf

3. The Fashion Show at the End of the World

https://unfccc.int/sites/default/files/
    resource/Fashion%20Industry%20
    Carter%20for%20Climate%20
    Action_2021.pdf
https://www.researchgate.net/
    profile/Patsy-Perry/publication/
    340635670_The_environmental_
    price_of_fast_fashion/links/
    5f2960c4a6fdcccc43a8ca65/The-
    environmental-price-of-fast-fashion.
    pdf
https://issuu.com/fashionrevolution/
    docs/fashiontransparencyindex_2021
https://www.sustainablejungle.com/
    sustainable-fashion/what-is-modal-
    fabric/#item%205
https://www.lenzing.com/
https://www.mckinsey.com/~/media/
    mckinsey/industries/retail/our%20

insights/fashion%20on%20climate/
    fashion-on-climate-full-report.pdf
https://www.weforum.org/agenda/
    2020/11/sustainable-fashion-reduce-
    greenhouse-gas-emissions
https://www.sciencedirect.com/topics/
    earth-and-planetary-sciences/
    synthetic-fiber
https://www.sustainyourstyle.org/old-
    environmental-impacts

4. Your Cat Doesn't Need to Eat Fish

https://globalfishingwatch.org/
    commercial-fishing/
https://www.fao.org/publications/sofia/
    en/
https://blog.nationalgeographic.org/
    2018/06/06/most-fishing-on-the-
    high-seas-would-be-unprofitable-at-
    current-rates-without-government-
    subsidies-a-new-study-reveals/
https://www.frontiersin.org/articles/
    10.3389/fmars.2019.00289/full

5. Your Fresh Air Is Asphyxiating You

https://projects.iq.harvard.edu/covid-
    pm
https://www.science.org/doi/10.1126/
    sciadv.abd4049
https://academic.oup.com/
    cardiovascres/article/116/14/2247/
    5940460
https://www.theguardian.com/
    environment/2020/nov/04/tiny-air-
    pollution-rise-linked-to-11-more-
    covid-19-deaths-study
https://www.science.org/doi/10.1126/
    sciadv.abd4049
https://www.sustainability-times.
    com/environmental-protection/

even-ancient-romans-badly-polluted-
europes-air/

https://www.who.int/news/item/02-05-
2018-9-out-of-10-people-worldwide-
breathe-polluted-air-but-more-
countries-are-taking-action

7. Hydrogen Makes Up 70 Per Cent of the Universe; I Didn't Know That Either

https://www2.lbl.gov/abc/wallchart/
chapters/10/0.html#toc

https://davidson.weizmann.ac.il/
en/online/orderoutofchaos/
body%E2%80%99s-elements

https://corporateeurope.org/sites/
default/files/2020-12/hydrogen-
report-web-final_3.pdf

8. Nuclear Power Is So Over

https://morningconsult.com/2020/09/
09/nuclear-energy-polling/

https://www.iisd.org/gsi/subsidy-
watch-blog/how-much-again-cost-
all-subsidies-hinkley-point-c-nuclear-
power-plant-should-be

https://www.ft.com/content/faba7e8a-
1983-4b47-9b70-ec903351a373

https://www.earthtrack.net/document/
nuclear-power-still-not-viable-
without-subsidies

https://www.cleantech.com/nuclear-
fusion-the-race-to-gain/

https://scitechdaily.com/scientists-
shatter-record-for-the-amount-of-
energy-produced-during-a-controlled-
sustained-fusion-reaction/

9. Stinky Gas

https://www.climatepolicyinitiative.
org/publication/global-landscape-of-
climate-finance-2021/

https://www.willistowerswatson.com/
en-GB/News/2020/10/global-asset-
manager-aum-tops-us-dollar-100-
trillion-for-the-first-time

https://influencemap.org/report/
Climate-Funds-Are-They-Paris-
Aligned-3eb83347267949847084306d
ae01c7b0

https://www.gov.ie/en/publication/
5052a-national-retrofit-plan/

https://www.irishtimes.com/news/
ireland/irish-news/q-a-how-will-the-
new-home-retrofitting-grants-work-
1.4797139

https://www.acer.europa.eu/gas-
factsheet

https://www.nrdc.org/experts/pierre-
delforge/gas-appliances-pollute-
indoor-and-outdoor-air-study-shows

https://www.commondreams.org/news/
2022/03/10/mckibben-heat-pumps-
peace-plan-gains-traction-biden

https://www.euractiv.com/section/
energy/news/industry-european-
electricity-grid-can-handle-50-
million-heat-pumps/

https://www.iea.org/reports/a-10-point-
plan-to-reduce-the-european-unions-
reliance-on-russian-natural-gas

10. Never Buy Carbon Offsets for Anything, Especially Your Car Gasoline

https://features.propublica.org/
brazil-carbon-offsets/inconvenient-
truth-carbon-credits-dont-work-
deforestation-redd-acre-cambodia/

https://www.co2.earth/co2-ice-core-data

https://www.sciencedirect.com/science/
article/pii/S2095809917303077

https://yaleclimateconnections.org/
2012/12/forget-about-that-2-degree-
future/

11. Please Don't Plant Trees

https://www.sciencedirect.com/science/article/abs/pii/S026499931630709X

https://www.irena.org/-/media/Files/IRENA/Agency/Publication/2021/Oct/IRENA_RE_Jobs_2021.pdf

https://www.theguardian.com/business/2019/oct/15/bank-of-england-boss-warns-global-finance-it-is-funding-climate-crisis

https://news.globallandscapesforum.org/49608/newly-seeded-with-14-billion-africas-great-green-wall-to-see-quicker-growth/

https://www.france24.com/en/live-news/20211123-back-in-the-spotlight-africa-s-great-green-wall

https://www.aljazeera.com/news/2021/10/26/green-or-greenwashing-saudi-arabias-climate-change-pledges

https://www.france24.com/en/live-news/20210327-saudi-arabia-unveils-campaign-to-tackle-climate-change

https://www.greenpeace.org.uk/news/the-biggest-problem-with-carbon-offsetting-is-that-it-doesnt-really-work/

12. Have as Many Babies as You Like

https://fairstartmovement.org/wp-content/uploads/2021/10/Art-16-Filing-Oct-2021.pdf

http://climatecasechart.com/climate-change-litigation/non-us-case/neubauer-et-al-v-germany/

https://www.climatechangenews.com/2021/05/05/germany-raises-ambition-net-zero-2045-landmark-court-ruling/

https://www.nytimes.com/2021/04/29/world/europe/germany-high-court-climate-change-youth.html

https://ican.adani.com/

https://reneweconomy.com.au/adani-says-it-wants-kids-to-take-climate-action-while-it-grows-fossil-fuel-empire/

https://www.census.gov/library/stories/2021/12/us-population-grew-in-2021-slowest-rate-since-founding-of-the-nation.html

13. Ride a Bicycle, Save the World

https://www.statista.com/statistics/200002/international-car-sales-since-1990/

https://www.bloomberg.com/news/features/2019-02-28/this-is-what-peak-car-looks-like

https://voxeu.org/article/potential-economic-and-social-effects-driverless-cars

https://www.thejakartapost.com/academia/2020/08/01/amid-pandemic-bike-boom-invest-in-wheels-of-change.html

14. Fly Without Guilt

https://www.tandfonline.com/doi/full/10.1080/09644016.2020.1863703

https://www.theatlantic.com/science/archive/2020/02/oil-industry-fighting-climate-policy-states/606640/

https://www.theguardian.com/tv-and-radio/2022/apr/20/what-we-now-know-they-lied-how-big-oil-companies-betrayed-us-all

https://www.theglobaleconomy.com/rankings/jet_fuel_production/Europe/

Energy Charter Treaty: https://www.tjm.org.uk/trade-deals/energy-charter-treaty

https://www.clientearth.org/latest/press-office/press/abandon-energy-charter-treaty-or-miss-climate-goals-lawyers-warn-commission/

https://www.energycharter.org/process/
energy-charter-treaty-1994/energy-
charter-treaty/
https://www.clientearth.org/projects/
the-greenwashing-files/rwe/
https://www.bloomberg.com/news/
articles/2020-12-13/nordic-nations-
set-pace-in-electric-planes-after-
green-cars-push
https://www.cnbc.com/2021/08/03/dhl-
express-buys-eviation-electric-planes-
for-us-package-delivery.html

15. A Luxury Cruise Liner Is a Stinking Floating Dumpster

https://e360.yale.edu/features/at-
last-the-shipping-industry-begins-
cleaning-up-its-dirty-fuels
https://airqualitynews.com/2019/06/
05/46-cruise-ships-emitted-10-times-
more-sulphur-than-260m-cars/
https://www.transportenvironment.org/
discover/one-corporation-pollute-
them-all/
https://www.transportenvironment.org/
discover/shipping-and-aviation-are-
subject-to-the-paris-agreement-legal-
analysis-shows/
https://inews.co.uk/news/long-reads/
cargo-container-shipping-carbon-
pollution-114721
https://www.cadmatic.com/en/
resources/articles/does-one-ship-
pollute-as-much-as-50-million-cars/
https://1bps6437gg8c169i0y1drtgz-
wpengine.netdna-ssl.com/wp-
content/uploads/2021/07/Cruise-
_Report_2021_v5.pdf
https://billmckibben.substack.com/p/
the-happiest-number-ive-heard-in?s=r
https://www.oilandgas360.com/loud-
calls-for-global-shipping-to-ditch-
fossil-fuels-and-meet-climate-goals/

https://theicct.org/silent-but-deadly-
the-case-of-shipping-emissions/
https://oceanconservancy.org/wp-
content/uploads/2020/09/Zero-
carbon-for-shipping_final.pdf
https://www.spglobal.com/commodity-
insights/en/market-insights/
latest-news/electric-power/020722-
uk-eyes-increased-on-shore-power-
for-shipping-at-british-ports-amid-
significant-cost-challenges
https://www.epa.gov/ports-initiative/
shore-power-technology-assessment-
us-ports
https://www.gminsights.com/industry-
analysis/gas-pipeline-infrastructure-
market

16. The Nasty Ninety

https://climateaccountability.org/
carbonmajors.html
https://link.springer.com/article/
10.1007/s10584-013-0986-y#Abs1
https://www.theguardian.com/
environment/2010/feb/18/worlds-top-
firms-environmental-damage
https://www.unepfi.org/news/putting-
a-price-on-global-environmental-
damage-news/
https://www.researchgate.net/
publication/288421148_AIDS_drugs_
for_all_social_movements_and_
market_transformations

17. The Social Media Axis of Evil

https://sciencepolicy.colorado.edu/
students/envs-geog_3022/ayling_2015.
pdf
https://theconversation.com/how-bill-
mckibbens-radical-idea-of-fossil-fuel-
divestment-transformed-the-climate-
debate-87895

https://gofossilfree.org/divestment/what-is-fossil-fuel-divestment/

18. Dial Down That Air Conditioning, But Not Too Much

https://japantoday.com/category/features/kuchikomi/debate-heats-up-over-cool-biz-temperature-rules

19. Going Vegan to Go Green? Don't Bother

https://www.ft.com/content/9b2466b7-1ef7-493d-83ea-a75a3687a7b6

https://news.gallup.com/poll/238328/snapshot-few-americans-vegetarian-vegan.aspx

https://www.reportbuyer.com/product/4959853/top-trends-in-prepared-foods-exploring-trends-in-meat-fish-and-seafood-pasta-noodles-and-rice-prepared-meals-savory-deli-food-soup-and-meat-substitutes.html

https://plantbasednews.org/culture/ethics/vegans-in-britain-skyrocketed/

https://www.nytimes.com/2019/09/30/health/red-meat-heart-cancer.html

https://www.technologyreview.com/2021/04/26/1023636/sustainable-meat-livestock-production-climate-change/

https://www.weforum.org/agenda/2021/07/2c-global-warming-difference-explained/

https://climate.nasa.gov/news/2865/a-degree-of-concern-why-global-temperatures-matter/

https://www.cell.com/one-earth/fulltext/S2590-3322(21)00233-5

https://www.theguardian.com/environment/2021/nov/18/the-forgotten-oil-ads-that-told-us-climate-change-was-nothing

https://www.nytimes.com/interactive/2019/12/12/climate/texas-methane-super-emitters.html

https://www.nytimes.com/2022/03/24/climate/methane-leaks-new-mexico.html

https://www.cnbc.com/2022/01/23/lab-grown-meat-start-ups-hope-to-make-strides-in-2022.html

https://www.statista.com/statistics/237597/leading-10-countries-worldwide-in-poultry-meat-production-in-2007/

https://unfccc.int/climate-action/momentum-for-change/planetary-health/impossible-foods

20. Drive Electric Shamelessly – the Green Energy Revolution Is Here

https://cleantechnica.com/2022/04/24/china-electric-car-market-reaches-26-plugin-market-share-in-march/

https://www.pv-magazine-australia.com/2022/04/27/full-extent-of-sun-cable-megaproject-revealed/

https://www.bloomberg.com/news/articles/2021-03-23/ev-charging-data-shows-a-widely-divergent-global-path?sref=En3bILwz

https://www.reuters.com/business/autos-transportation/exclusive-global-carmakers-now-target-515-billion-evs-batteries-2021-11-10/

https://www.theguardian.com/environment/2019/oct/10/exclusive-carmakers-opponents-climate-action-us-europe-emissions and https://mitsloan.mit.edu/ideas-made-to-matter/calculating-damage-automakers-pollution-collusion and https://www.tse-fr.eu/publications/colluding-against-environmental-regulation

284

21. Green Bonds Do More Harm Than Good

https://www.climatebonds.net/2021/
09/greenium-remains-visible-latest-
pricing-study

https://www.gov.pl/web/finance/issues-
international-bonds

22. Tinker, Lawyer, Banker, Fry

https://www.jpmorgan.com/solutions/
cib/investment-banking/center-for-
carbon-transition/carbon-compass

https://reclaimfinance.org/site/en/
2021/09/23/banks-and-investors-
pouring-billions-into-arctic-oil-gas-
bonanzadespite-climate-pledges-
report/

https://www.banktrack.org/campaign/
banks_and_fossil_fuel_financing_
and_funding https://reclaimfinance.
org/site/en/2021/03/24/baking-
climate-chaos-fossil-fuel-finance-
report-2021/

https://insure-our-future.com/wp-
content/uploads/2021/11/2021-IOF-
Scorecard.pdf

https://www.ls4ca.org/ https://
static1.squarespace.com/static/
5f53fa556b708446acb4dcb5/t/
611dba29c5ad3077663d4947/
1629338162366/2021+Law+Firm+Cli
mate+Change+Scorecard.pdf

https://www.iea.org/reports/net-zero-
by-2050

https://www.theguardian.com/
environment/2021/may/18/no-new-
investment-in-fossil-fuels-demands-
top-energy-economist

23. The ESG Con

https://insideclimatenews.org/book/
exxon-the-road-not-taken/

https://influencemap.org/report/
Climate-Funds-Are-They-Paris-
Aligned-3eb83347267949847084306d
ae01c7b0

24. Don't Worry (At All) About Bitcoin's Energy Use

https://ccaf.io/cbeci/index
https://ourworldindata.org/energy-
production-consumption
https://ccaf.io/cbeci/index/comparisons

25. Love Thy Insect

https://wwf.panda.org/discover/our_
focus/biodiversity/biodiversity/

https://www.globaldealfornature.org/
science/

https://www.cbd.int/convention/

https://www.oecd.org/environment/
resources/biodiversity/report-a-
comprehensive-overview-of-global-
biodiversity-finance.pdf

https://www.science.org/doi/10.1126/
sciadv.aaw2869 https://www.
globaldealfornature.org/science/

https://www.paulsoninstitute.org/
conservation/financing-nature-report/

https://www.statista.com/outlook/cmo/
footwear/worldwide#revenue

https://inews.co.uk/news/environment/
honey-bees-disease-climate-change-
rise-uk-yorkshire-norfolk-south-west-
areas-1291422

https://ocm.auburn.edu/newsroom/
news_articles/2019/02/251601-bee-
pollination-market.php

https://news.globallandscapesforum.
org/54916/the-amazon-rainforest-is-
nearing-its-tipping-point-but-what-
does-that-mean/

26. The Royal Baby Versus Biodiversity

https://ipbes.net/global-assessment
https://pressgazette.co.uk/press-coverage-of-climate-change-analysed-by-country-uk-and-australia-rank-top/
http://sciencepolicy.colorado.edu/icecaps/research/media_coverage/index.html
https://www.highnorthnews.com/en/cargo-volume-northern-sea-route-reaches-35m-tons-record-number-transits

27. Sue the Bastards

https://www.cnbc.com/2021/11/11/cop26-climate-campaigners-to-target-banks-after-shell-court-ruling.html
https://youth4climatejustice.org/
https://www.jonesday.com/en/insights/2021/06/climate-change-litigation-bombshell-dutch-lower-court-orders-royal-dutch-shell-to-reduce-co2-emissions
https://www.ft.com/content/d7feaa8a-7555-47ae-828b-274527c6f89c
https://www.dawn.com/news/1269246
https://aida-americas.org/en/press/new-law-banning-mining-colombia-s-p-ramos-could-draw-its-first-lawsuit

28. It's Raining Renewable Energy

https://www.iea.org/reports/world-energy-investment-2021/executive-summary
https://www.inet.ox.ac.uk/files/energy_transition_paper-INET-working-paper.pdf
https://www.reuters.com/markets/commodities/oil-firms-face-workforce-crunch-renewables-beckon-survey-2021-11-30/
https://ember-climate.org/insights/research/global-electricity-review-2022/

Epilogue

https://www.nytimes.com/2005/07/19/world/asia/anger-in-china-rises-over-threat-to-environment.html
https://www.nytimes.com/2018/03/13/world/asia/china-xi-jinping-congress-pollution-corruption.html
https://thediplomat.com/2019/07/environmental-protest-breaks-out-in-chinas-wuhan-city/
https://e360.yale.edu/features/as-it-looks-to-go-green-china-keeps-a-tight-lid-on-dissent

Index

Ostend, Belgium, 65–6
OVS (fashion brand), 31
Oxford University, 268–70, 272
ozone, 49, 142, 162

P&O Cruises, 137
Pakistan, 108, 257–9
palm oil, 21–7
Paris, France, 119–20, 124
Paris Agreement (2015), 5, 8, 27, 44–5,
    56, 82, 139, 143, 201–3, 220, 222
Patagonia (fashion brand), 31
Paulson Institute, 248
Pemex, 68
pension funds, 148–9
PepsiCo, 11, 24, 195
permafrost, 253–4
Peru, 259
pesticides, 243
Petrobras, 68
petrochemical companies, 11, 14–19
PetroChina, 172
pharmaceutical companies, 144–5
Philippines, 259
Plastic China (film), 275
plastics, 9–20, 30, 37, 227–8, 250
  bioplastics, 12
  microplastics, 9–10, 28–30, 37
  recycling, 13–14, 18–19, 117
Poland, 143, 198
pollution, 1–3, 94–6, 106, 108, 131–2,
    136–42, 193, 203, 228, 268
  air pollution, 23, 47–52, 119–24, 125,
    142, 190, 275
  methane pollution, 173–4, 176, 178
  plastic pollution, 12–20, 30, 195, 250
  shipping pollution, 136–42
  water pollution, 142, 190
Polman, Paul, 195
polyester, 29, 34
polymers, 9, 11
population, 113–18
Population Balance, 113
Portugal, 181, 257
Prada, 28, 31

Premier Oil, 216
Princess Cruises, 137
Procter & Gamble, 11
Puma, 37

Quiksilver, 31

Rainforest Trust, 247
Ralph Lauren, 36–7
RE 100 initiative, 35–7
Reclaim Finance, 205
Redwood Materials, 189
Reebok, 31
renewable energy, 35–6, 66–7, 80,
    89–90, 109–10, 141, 164–5, 183–90,
    238, 263–73
  see also hydro power; hydrogen; solar
    power; tidal power; wind power
Repsol, 68
Re:wild, 247
Rifkin, Jeremy, 61
Roadmap to Zero Programme, 33
Rob and Melani Walton Foundation,
    247
Rockhopper, 131
Rolling Stone magazine, 158
Rosneft, 172
Rowland, F. Sherwood, 162
Royal Bank of Canada, 197
Russia, 86, 90, 108, 143, 254
Rwanda, 17–18
RWE, 130–31, 156–7

Saks Fifth Avenue, 31
SARS virus, 252
Saudi Arabia, 105, 235
Saudi Aramco, 15, 17, 56, 68, 105, 143,
    152, 172
Science Advances, 48
Science Based Targets Initiative, 32, 58
Scotland, 267
Seabourn, 137
Seaspiracy (documentary), 41
Securities and Exchange Commission
    (SEC), 223

Intergovernmental Science-Policy
  Platform on Biodiversity and
  Ecosystem Services (IPBES), 249–51
  renewable energy and, 186
  Strategic Plan for Biodiversity, 255
  Treaty on Business and Human Rights,
  25
United States, 14–16, 42, 47–8, 59, 108,
  204, 224, 258
  nuclear power and, 71–2, 74
  natural gas and, 87, 171–2
  climate change and, 94–5, 201–2
  cycling, 121–2
  air conditioning, 160–61
  renewable energy and, 236–7, 266, 272
UPS, 133
urban planning, see cities
US Business Roundtable, 55

Vans, 31
veganism/vegetarianism, 167–79
Vehicle-to-grid technology (V2G), 182
Venezuela, 232, 235
Vestas, 188
Vietnam, 271
viruses, 51–2, 240, 252–3
  see also coronavirus
Vitol, 265
Volkswagen, 181–2
Volvo, 181

water, 142
  shortages, 30

wastewater, 30
  see also hydro power; tidal power
We Mean Business Coalition, 58
Wells Fargo, 16, 204
wetlands, 252
wildlife, 244–5
  animal pathogens, 51–2, 252
  endangered species, 23, 243
  plastic pollution and, 12–13
  see also biodiversity; insects
Wildlife Trust, 252
wind power, 66, 70, 77, 183–5, 263,
  271–2
  offshore wind farms, 66, 89, 266, 271
  wind turbine recycling, 187–8
working from home, 125
World Bank, 58, 83, 107
World Trade Organization (WTO) 46,
  50
World Wide Fund for Nature, 58, 244
World Wildlife Fund, 57
Wright's Law, 270–72
Wyss Foundation, 247

Xi Jinping, 275

Yucca Mountain Nuclear Waste
  Repository, Nevada, 71–2

Zara, 28
ZeroAvia, 134
Zoological Society of London, 252